I0585018

John Fiske

A Century of Science, and Other Essays

John Fiske

A Century of Science, and Other Essays

ISBN/EAN: 9783337034764

Printed in Europe, USA, Canada, Australia, Japan

Cover: Foto ©berggeist007 / pixelio.de

More available books at **www.hansebooks.com**

By Mr. Fiske.

ESSAYS AND PHILOSOPHY.

A CENTURY OF SCIENCE, and other Essays. Crown 8vo, $2.00.

MYTHS AND MYTH-MAKERS: Old Tales and Superstitions interpreted by Comparative Mythology. Crown 8vo, $2.00.

OUTLINES OF COSMIC PHILOSOPHY. Based on the Doctrine of Evolution, with Criticisms on the Positive Philosophy. 2 vols. crown 8vo, $6.00.

THE UNSEEN WORLD, and other Essays. Crown 8vo, $2.00.

EXCURSIONS OF AN EVOLUTIONIST. Crown 8vo, $2.00.

DARWINISM, and other Essays. Crown 8vo, $2.00.

THE DESTINY OF MAN, viewed in the Light of His Origin. 16mo, $1.00.

THE IDEA OF GOD, as affected by Modern Knowledge. 16mo, $1.00.

THROUGH NATURE TO GOD. 16mo, $1.00.

For complete list of Mr. Fiske's Historical and Philosophical Works and Essays, see pages at the back of this work.

HOUGHTON, MIFFLIN & COMPANY,
BOSTON AND NEW YORK.

A CENTURY OF SCIENCE

And Other Essays

BY

JOHN FISKE

Out of the shadows of night
The world rolls into light:
It is daybreak everywhere.
LONGFELLOW.

The Riverside Press

BOSTON AND NEW YORK
HOUGHTON, MIFFLIN AND COMPANY
The Riverside Press, Cambridge
1899

COPYRIGHT, 1899, BY JOHN FISKE

ALL RIGHTS RESERVED

DEDICATORY EPISTLE

TO

THOMAS SERGEANT PERRY,

PROFESSOR OF ENGLISH LITERATURE IN THE KEIO GIJUKU, AT TOKYO.

DEAR TOM, — It has long been my wish to make
you the patron saint or tutelar divinity of some
book of mine, and it has lately occurred to me that
it ought to be a book of the desultory and chatty
sort that would remind you, in 'your present exile
at the world's eastern rim, of the many quiet evenings of old, when, over a tankard of mellow October and pipe of fragrant Virginia, while Yule logs
crackled blithely and the music of pattering sleet
was upon the window-pane, we used to roam in
fancy through the universe and give free utterance
to such thoughts, sedate or frivolous, as seemed to
us good. I dare say the present volume may serve
as an epitome of many such old-time sessions of
sweet discourse, which I trust we shall by and by
repeat and renew.

But there is one link of association which in my
mind especially connects you with the present occa-

sion.   My theory of the causes and effects of the
prolongation of human infancy, with reference to
the evolution of man, was first published in the
"North American Review" for October, 1873,
when you were the editor of that periodical.   The
article, which was entitled "The Progress from
Brute to Man," was made up of two chapters of
my "Outlines of Cosmic Philosophy" (part ii.
chaps. xxi., xxii.), which was published a year
later, in October, 1874.   The value of the theory
therein set forth was at once recognized by many
leading naturalists.   In the address of Vice-Presi-
dent Edward Morse, before the American Association, at its meeting at Buffalo in 1876, my theory
receives extended notice as one of the most important contributions yet made to the Doctrine of Evolution ; and it is declared that I have given " for
the first time a rational explanation of the origin
and persistence of family relations, and thence
communal [*i. e.*, clan] relations, and, finally, of
society." [1]

Uncontrollable circumstances have prevented
my giving to the further elaboration of this infancy
theory the time and attention which it deserves

----

[1] Morse, *What American Zoölogists have done for Evolution*,
pp. 37, 39–41, Salem, 1876; *Proc. Amer. Assoc. for Adv. of Sci.*,
vol. xxii.

and demands; but in my little book, " The Destiny
of Man," published in 1884, I gave a popular exposition of it which has made it widely known in
all English-speaking countries and on the continent
of Europe, as well as among your worthy Japanese
neighbours, Tom, who have done me the honour to
translate some of my books into their vernacular.
The theory has become still further popularized
through having furnished the starting-point for
some of the most characteristic speculations of the
late Henry Drummond. In these and other ways
my infancy theory has so far entered into the current thoughts of the present age that people have
(naturally enough) begun to forget with whom it
originated. For example, in the recent book,
" Through Nature to God," while criticising a remark of Huxley's, I found it desirable to make a
restatement of the infancy theory; whereupon a
friendly reviewer, referring to that particular part
of the book, observes that "of course" it makes no
pretensions to originality, but is simply my lucid
summary of speculations with which every reader
of Darwin, Spencer, Huxley, Romanes, and Drummond is familiar ! In point of fact, not the faintest
suggestion of this infancy theory can be found in
all the writings of Darwin, Huxley, and Romanes.
In Spencer's "Sociology," vol. i. p. 630, it is briefly

mentioned with approval as an important contribution originating with me ; and in Drummond's "Ascent of Man," which is really built upon it, credit is cordially given me.[1]

Indeed, down to the present time, I have been left almost in exclusive possession of that area of speculation which is occupied with the genesis of Man as connected with that prolongation of infancy which first began to become conspicuous in the manlike apes. There are many who assent to what I have put forth, but few who seem inclined to enter that difficult field on the marchland between biology, psychology, and sociology. Doubtless this is because the attention of the scientific world has for forty years been absorbed in the more general questions concerning the competency of natural selection, the causes of variation, the agencies alleged by Lamarck, and in these latter days Weismannism, etc. In course of time, however, the more special problems connected with man's genesis will surely come uppermost, and then we may hope to see the causes of the lengthening of infancy investigated by thinkers duly conversant alike with psychology and embryology.

Questions of priority in originating new theories

[1] *The Ascent of Man*, pp. 282-291; cf. Tyler, *The Whence and the Whither of Man*, pp. 179, 217, etc.

may not greatly interest the general reader, but you and I feel interested in preventing any misconception in the present case; and it was thus that the careless remark of the friendly reviewer led me to insert in the present volume the shorthand report of some autobiographical remarks on the infancy theory. In reading the proof-sheets I have noticed that the book contains elsewhere many allusions to personal experiences. This feature, which was quite unforeseen, will not fail to commend it all the more strongly to you, my ancient friend and comrade. As for readers in general, I may best conclude in the words of old Aaron Rathbone, whose book entitled "The Surveyor" was dated "from my lodging at the house of M. Roger Bvrgis, against Salisburie-house-gate, in the Strand, this sixt of Nouember, 1616." This wise and placid philosopher saith: "To perswade the courteous were causelesse, for they are naturally kind; and to diswade the captious were bootless, for they will not be diverted. Let the first make true vse of these my labours, and they shall find pleasure and profit therein; let the last (if they like not) leave it, and it shall not offend them."

Wherefore let me, without further ado, subscribe myself,   Ever yours,

JOHN FISKE.

CAMBRIDGE, *October* 25, 1899.

CONTENTS

A CENTURY OF SCIENCE

I

A CENTURY OF SCIENCE [1]

In the course of the year 1774 Dr. Priestley found that by heating red precipitate, or what we now call red oxide of mercury, a gas was obtained, which he called "dephlogisticated air," or, in other words, air deprived of phlogiston, and therefore incombustible. This incombustible air was *oxygen*, and such was man's first introduction to the mighty element that makes one fifth of the atmosphere in volume and eight ninths of the ocean by weight, besides forming one half of the earth's solid crust, and supporting all fire and all life. I know of nothing which can reveal to us with such startling vividness the extent of the gulf which the human mind has traversed within little more than a hundred years. It is scarcely possible to put ourselves back into the frame of mind

[1] An address delivered in the First Unitarian Church of Philadelphia, May 13, 1896, at the celebration of the one hundredth anniversary of its founding, under the lead of the illustrious Dr. Priestley.

in which oxygen was unknown, and no man could
tell what takes place when a log of wood is burned
on the hearth. The language employed by Dr.
Priestley carries us back to the time when chemistry was beginning to emerge from alchemy. It
was Newton's contemporary, Stahl, who invented
the doctrine of phlogiston in order to account for
combustion. Stahl supposed that all combustible
substances contain a common element, or fire principle, which he called phlogiston, and which escapes in the process of combustion. Indeed, the
act of combustion was supposed to consist in the
escape of phlogiston. Whither this mysterious fire
principle betook itself, after severing its connection
with visible matter, was not too clearly indicated,
but of course it was to that limbo far larger than
purgatory, the oubliette wherein have perished
men's unsuccessful guesses at truth. Stahl's theory, however, marked a great advance upon what
had gone before, inasmuch as it stated the case in
such a way as to admit of direct refutation. Little
use was made of the balance in those days, but
when it was observed that zinc and lead and sundry other substances grow heavier in burning, it
seemed hardly correct to suppose that anything
had escaped from these substances. To this objection the friends of the fire principle replied that

phlogiston might weigh less than nothing, or, in
other words, might be endowed with a positive attribute of levity, so that to subtract it from a
body would increase the weight of the body. This
was a truly shifty method of reasoning, in which
your phlogiston, with its plus sign to-day and its
minus sign to-morrow, exhibited a skill in facing
both ways like that of an American candidate for
public office.

Into the structure of false science that had been
reared upon these misconceptions Dr. Priestley's
discovery of oxygen came like a bombshell. As in
so many other like cases, the discovery was destined
to come at about that time ; it was made again
three years afterward by the Swedish chemist
Scheele, without knowing what Priestley had done.
The study of oxygen soon pointed to the conclusion
that, whatever may escape during combustion, oxygen is always united with the burning substance.
Then came Lavoisier with his balance, and proved
that whenever a thing burns it combines with
Priestley's oxygen, and the weight of the resulting
product is equal to the weight of the substance
burned plus the weight of oxygen abstracted from
the air. Thus combustion is simply union with
oxygen, and nothing escapes. No room was left
for phlogiston. Men's thoughts were dephlogis-

ticated from that time forth. The balance became the ruling instrument of chemistry. One further step led to the generalization that in all chemical changes there is no such thing as increase or diminution, but only substitution, and upon this fundamental truth of the indestructibility of matter all modern chemistry rests.

When we look at the stupendous edifice of science that has been reared upon this basis, when we consider the almost limitless sweep of inorganic and organic chemistry, the myriad applications to the arts, the depth to which we have been enabled to penetrate into the innermost proclivities of matter, it seems almost incredible that a single century can have witnessed so much achievement. We must admit the fact, but our minds cannot take it in; we are staggered by it. One thing stands out prominently, as we contrast this rapid and coherent progress with the barrenness of ancient alchemy and the chaotic fumbling of the Stahl period: we see the importance of untrammelled inquiry, and of sound methods of investigation which admit of verification at every step. That humble instrument the balance, working in the service of sovereign law, has been a beneficent Jinni unlocking the portals of many a chamber wherein may be heard the secret harmonies of the world.

It is not only in chemistry, however, that the marvellous advance of science has been exhibited. In all directions the quantity of achievement has been so marked that it is worth our while to take a brief general survey of the whole, to see if haply we may seize upon the fundamental characteristics of this great progress. In the first place, a glance at astronomy will show us how much our knowledge of the world has enlarged in space since the day when Priestley set free his dephlogisticated air.

The known solar system then consisted of sun, moon, earth, and the five planets visible to the naked eye. Since the days of the Chaldæan shepherds there had been no additions except the moons of Jupiter and Saturn. Herschel's telescope was to win its first triumph in the detection of Uranus in 1781. The Newtonian theory, promulgated in 1687, had come to be generally accepted, but there were difficulties remaining, connected with the planetary perturbations and the inequalities in the moon's motion, which the glorious labours of Lagrange and Laplace were presently to explain and remove, — labours which bore their full fruition two generations later, in 1845, when the discovery of the planet Neptune, by purely mathematical reasoning from the observed effects of its gravitation, furnished for the Newtonian theory the grandest confirmation known in the whole history of science. In Priestley's time, sidereal astronomy was little more than the cataloguing of such stars and nebulæ as could be seen with the telescopes then at command. Sixty years after the discovery of oxygen the distance of no star had been measured. In 1836, Auguste Comte assured his readers that such a feat was impossible, that the Newtonian theory could never be proved to extend through the interstellar spaces, and that the matter of which stars are composed may be entirely different in its properties from the matter with which we are familiar. Within three years the first part of this prophecy was disproved when Bessel measured the distance of the star 61 Cygni; since then the study of the movements of double and multiple stars has shown them conforming to Newton's law; and as for the matter of which they are composed, we are introduced to a chapter in science which even the boldest speculator of half a century ago would have derided as a baseless dream. The discovery of spectrum analysis and the invention of the spectroscope, completed in 1861 by Kirchhoff and Bunsen, have supplied data for the creation of a stellar chemistry; showing us, for example, hydrogen in Sirius and the

nebula of Orion, sodium and potassium, calcium
and iron, in the sun ; demonstrating the gaseous
character of nebulæ ; and revealing chemical elements hitherto unknown, such as helium, a mineral
first detected in the sun's atmosphere, and afterward found in Norway. A still more wonderful
result of spectrum analysis is our ability to measure the motion of a star through a slight shifting
in the wave-lengths of the light which it emits.
In this way we can measure, in the absence of all
parallax, the direct approach or recession of a star ;
and in somewhat similar wise has been discovered
the cause of the long-observed variations of brilliancy in Algol. That star, which is about the
size of our sun, has a dark companion not much
smaller, and the twain are moving around a third
body, also dark : the result is an irregular series
of eclipses of Algol, and the gravitative forces exerted by the two invisible stars are estimated
through their effects upon the spectrum of the
bright star. In no department of science has a
region of inference been reached more remote than
this. From such a flight one may come back
gently to more familiar regions while remarking
upon the manifold results that have begun to be
attained from the application of a sensitive photo-
graph plate to the telescope in place of the human

eye. It may suffice to observe that we thus catch the fleeting aspects of sun-spots and preserve them for study; we detect the feeble self-luminosity still left in such a slowly cooling planet as Jupiter; and since the metallic plate does not quickly weary, like the human retina, the cumulative effects of its long exposure reveal the existence of countless stars and nebulæ too remote to be otherwise reached by any visual process. By such photographic methods George Darwin has caught an equatorial ring in the act of detachment from its parent nebula, and the successive phases of the slow process may be watched and recorded by generations of mortals yet to come.

To appreciate the philosophic bearings of this vast enlargement of the mental horizon, let us recall just what happened when Newton first took the leap from earth into the celestial spaces by establishing a law of physics to which moon and apple alike conform. It was the first step, and a very long one, toward proving that the terrestrial and celestial worlds are dynamically akin, that the same kind of order prevails through both alike, that both are parts of one cosmic whole. So late as Kepler's time, it was possible to argue that the planets are propelled in their elliptic orbits by forces quite unlike any that are disclosed by purely terrestrial experience, and therefore perhaps inaccessible to any rational interpretation. Such imaginary lines of demarcation between earth and heavens were forever swept away by Newton, and the recent work of spectrum analysis simply completes the demonstration that the remotest bodies which the photographic telescope can disclose are truly part and parcel of the dynamic world in which we live.

All this enlargement of the mental horizon, from Newton to Kirchhoff, had reference to space. The nineteenth century has witnessed an equally notable enlargement with reference to time. The beginnings of scientific geology were much later than those of astronomy. The phenomena were less striking and far more complicated ; it took longer, therefore, to bring men's minds to bear upon them. Antagonism on the part of theologians was also slower in dying out. The complaint against Newton, that he substituted Blind Gravitation for an Intelligent Deity, was nothing compared to the abuse that was afterwards lavished upon geologists for disturbing the accepted Biblical chronology. At the time when Priestley discovered oxygen, educated men were still to be found who could maintain with a sober face that fossils had been created already dead and petrified, just for the fun of the thing. The writings of Buffon were preparing men's minds for the belief that the earth's crust has witnessed many and important changes, but there could be no scientific geology until further progress was made in physics and chemistry. It was only in 1763 that Joseph Black discovered latent heat, and thus gave us a clue to what happens when water freezes and melts, or when it is turned into steam. It was in 1786 that the publication of James Hutton's "Theory of the Earth" ushered in the great battle between Neptunians and Plutonists which prepared the way for scientific geology. When the new science won its first great triumph with Lyell in 1830, the philosophic purport of the event was the same that was being proclaimed by the progress of astronomy. Newton proved that the forces which keep the planets in their orbits are not strange or supernatural forces, but just such as we see in operation upon this earth every moment of our lives. Geologists before Lyell had been led to the conclusion that the general aspect of the earth's surface with which we are familiar is by no means its primitive or its permanent aspect, but that there has been a succession of ages, in which the relations of land and water, of mountain and plain, have varied to a very considerable extent; in which soils and climates have undergone most complicated vicissitudes; and in which the earth's vegetable products and its animal populations have again and again assumed new forms, while the old forms have passed away. In order to account for such wholesale changes, geologists were at first disposed to imagine violent catastrophes brought about by strange agencies, — agencies which were perhaps not exactly supernatural, but were in some vague, unspecified way different from those which are now at work in the visible and familiar order of nature. But Lyell proved that the very same kind of physical processes which are now going on about us would suffice, during a long period of time, to produce the changes in the inorganic world which distinguish one geological period from another. Here, in Lyell's geological investigations, there was for the first time due attention paid to the immense importance of the prolonged and cumulative action of slight and unobtrusive causes. The continual dropping that wears away stones might have served as a text for the whole series of beautiful researches of which he first summed up the results in 1830. As astronomy was steadily advancing toward the proof that in the abysses of space the physical forces at work are the same as our terrestrial forces, so geology, in carrying us

back to enormously remote periods of time, began
to teach that the forces at work have all along
been the same forces that are operative now.   Of
course, in that early stage when the earth's crust
was in process of formation, when the temperature
was excessively high, there were phenomena here
such as can no longer be witnessed, but for which
we must look to big planets like Jupiter ; in that
intensely hot atmosphere violent disturbances occur, and chemical elements are dissociated which we
are accustomed to find in close combination here.
But ever since our earth cooled to a point at which
its solid crust acquired stability, since the earliest
mollusks and vertebrates began to swim in the
seas and worms to crawl in the damp ground, if at
almost any time we could have come here on a
visit, we should doubtless have found things going
on at measured pace very much as at present, —
here and there earthquake and avalanche, fire and
flood, but generally rain falling, sunshine quick-
ening, herbage sprouting, creatures of some sort
browsing, all as quiet and peaceful as a daisied
field in June, without the slightest visible presage
of the continuous series of minute secular changes
that were gradually to transform a Carboniferous
world into what was by and by to be a Jurassic
world, and that again into what was after a while

to be an Eocene world, and so on, until the aspect of the world that we know to-day should noiselessly steal upon us.

When once the truth of Lyell's conclusions began to be distinctly realized, their influence upon men's habits of thought and upon the drift of philosophic speculation was profound. The conception of Evolution was irresistibly forced upon men's attention. It was proved beyond question that the world was not created in the form in which we find it to-day, but has gone through many phases, of which the later are very different from the earlier; and it was shown that, so far as the inorganic world is concerned, the changes can be much more satisfactorily explained by a reference to the ceaseless, all-pervading activity of gentle, unobtrusive causes such as we know than by an appeal to imaginary catastrophes such as we have no means of verifying. It began to appear, also, that the facts which form the subject-matter of different departments of science are not detached and independent groups of facts, but that all are intimately related one with another, and that all may be brought under contribution in illustrating the history of cosmic events. It was a sense of this interdependence of different departments that led Auguste Comte to write his " Philosophie Positive," the first volume of which appeared in 1830, in which he sought to point out the methods which each science has at command for discovering truth, and the manner in which each might be made to contribute toward a sound body of philosophic doctrine. The attempt had a charm and a stimulus for many minds, but failed by being enlisted in the service of sundry sociological vagaries upon which the author's mind was completely wrecked. " Positivism," from being the name of a potent scientific method, became the name of one more among the myriad ways of having a church and regulating the details of life.

While the ponderous mechanical intellect of Comte was striving to elicit the truth from themes beyond its grasp, one of the world's supreme poets had already discerned some of the deeper aspects of science presently to be set forth. By temperament and by training, Goethe was one of the first among evolutionists. The belief in an evolution of higher from lower organisms could not fail to be strongly suggested to a mind like his as soon as the classification of plants and animals had begun to be conducted upon scientific principles. It is not for nothing that a table of classes, orders, families, genera, and species, when graphically laid out, resembles a family tree. It was not long after

Linnæus that believers in some sort of a development theory, often fantastic enough, began to be
met with. The facts of morphology gave further
suggestions in the same direction. Such facts were
first generalized on a grand scale by Goethe in his
beautiful little essay on " The Metamorphoses of
Plants,"written in 1790, and his " Introduction to
Morphology," written in 1795, but not published
until 1807. In these profound treatises, which were
too far in advance of their age to exert much influence at first, Goethe laid the philosophic foundations
of comparative anatomy in both vegetal and animal
worlds. The conceptions of metamorphosis and of
homology, which were thus brought forward, tended
powerfully toward a recognition of the process of
evolution. It was shown that what under some
circumstances grows into a stem with a whorl of
leaves, under other circumstances grows into a
flower; it was shown that in the general scheme
of the vertebrate skeleton a pectoral fin, a fore leg,
and a wing occupy the same positions : thus was
strongly suggested the idea that what under some
circumstances developed into a fin might under
other circumstances develop into a leg or a wing.
The revelations of palæontology, showing various
extinct adult forms, with corresponding organs
in various degrees of development, went far to

strengthen this suggestion, until an unanswerable argument was reached with the study of rudimentary organs, which have no meaning except as remnants of a vanished past during which the organism has been changing. The study of comparative embryology pointed in the same direction; for it was soon observed that the embryos and larvæ of the higher forms of each group of animals pass, " in the course of their development, through a series of stages in which they more or less completely resemble the lower forms of the group." [1]

Before the full significance of such facts of embryology and morphology could be felt, it was necessary that the work of classification should be carried far beyond the point at which it had been left by Linnæus. In mapping out the relationships in the animal kingdom, the great Swedish naturalist had relied less than his predecessors upon external or superficial characteristics; the time was arriving when classification should be based upon a thorough study of internal structure, and this was done by a noble company of French anatomists, among whom Cuvier was chief. It was about 1817 that Cuvier's gigantic work reached its climax in bringing palæontology into alliance with systematic zoölogy, and effecting that

[1] Balfour, *Comparative Embryology*, i. 2.

grand classification of animals in space and time
which at once cast into the shade all that had gone
before it. During the past fifty years there have
been great changes made in Cuvier's classification,
especially in the case of the lower forms of animal
life. His class of *Radiata* has been broken up,
other divisions in his invertebrate world have been
modified beyond recognition, his vertebrate scheme
has been overhauled in many quarters, his attempt
to erect a distinct order for Man has been overthrown. Among the great anatomists concerned
in this work the greatest name is that of Huxley.
The classification most generally adopted to-day is
Huxley's, but it is rather a modification of Cuvier's
than a new development. So enduring has been
the work of the great Frenchman.

With Cuvier the analysis of the animal organism made some progress in such wise that anatomists began to concentrate their attention upon
the study of the development and characteristic
functions of organs. Philosophically, this was a
long step in advance, but a still longer one was
taken at about the same time by that astonishing
youth whose career has no parallel in the history
of science. When Xavier Bichat died in 1802, in
his thirty-first year, he left behind him a treatise
on comparative anatomy in which the subject was

worked up from the study of the tissues and their properties. The path thus broken by Bichat led to the cell doctrine of Schleiden and Schwann, matured about 1840, which remains, with some modifications, the basis of modern biology. The advance along these lines contributed signally to the advancement of embryology, which reached a startling height in 1829 with the publication of Baer's memorable treatise, in which the development of an ovum is shown to consist in a change from homogeneity to heterogeneity through successive differentiations. But while Baer thus arrived at the very threshold of the law of evolution, he was not in the true sense an evolutionist; he had nothing to say to phylogenetic evolution, or the derivation of the higher forms of life from lower forms through physical descent with modifications. Just so with Cuvier. When he effected his grand classification, he prepared the way most thoroughly for a general theory of evolution, but he always resisted any such inference from his work. He was building better than he knew.

The hesitancy of such men as Cuvier and Baer was no doubt due partly to the apparent absence of any true cause for physical modifications in species, partly to the completeness with which their own great work absorbed their minds. Often in the history of science we witness the spectacle of a brilliant discoverer travelling in triumph along some new path, but stopping just short of the goal which subsequent exploration has revealed. There it stands looming up before his face, but he is blind to its presence through the excess of light which he has already taken in. The intellectual effort already put forth has left no surplus for any further sweep of comprehension, so that further advance requires a fresher mind and a new start with faculties unjaded and unwarped. To discover a great truth usually requires a succession of thinkers. Among the eminent anatomists who in the earlier part of our century were occupied with the classification of animals, there were some who found themselves compelled to believe in phylogenetic evolution, although they could frame no satisfactory theory to account for it. The weight of evidence was already in favour of such evolution, and these men could not fail to see it. Foremost among them was Jean Baptiste Lamarck, whose work was of supreme importance. His views were stated in 1809 in his "Philosophie Zoölogique," and further illustrated in 1815, in his voluminous treatise on invertebrate animals. Lamarck entirely rejected the notion of special creations, and he pointed out some of the important factors in evo-

lution, especially the law that organs and faculties
tend to increase with exercise, and to diminish with
disuse.   His weakest point was the disposition to
imagine some inherent and ubiquitous tendency toward evolution, whereas a closer study of nature has
taught us that evolution occurs only where there
is a concurrence of favourable conditions.   Among
others who maintained some theory of evolution
were the two Geoffroy Saint-Hilaires, father and
son, and the two great botanists, Naudin in France
and Hooker in England.   In 1852 the case of evolution as against special creations was argued by
Herbert Spencer with convincing force, and in
1855 appeared " The Principles of Psychology,"
by the same author, a book which is from beginning to end an elaborate illustration of the process
of evolution, and is divided from everything that
came before it by a gulf as wide as that which
divides the Copernican astronomy from the Ptole-
maic.

The followers of Cuvier regarded the methods
and results of these evolutionists with strong dis-
approval.   In the excess of such a feeling, they
even went so far as to condemn all philosophic
thinking on subjects within the scope of natural
history as visionary and unscientific.   Why seek
for any especial significance in the fact that every

spider and every lobster is made up of just twenty
segments? Is it not enough to know the fact?
Children must not ask too many questions. It is
the business of science to gather facts, not to seek
for hidden implications. Such was the mental attitude into which men of science were quite commonly driven, between 1830 and 1860, by their desire to blink the question of evolution. A feeling
grew up that the true glory of a scientific career was
to detect for the two hundredth time an asteroid,
or to stick a pin through a beetle with a label at-
tached bearing your own latinized name, *Browni*,
or *Jonesii*, or *Robinsoniense*. This feeling was
especially strong in France, and was not confined
to physical science. It was exhibited a few years
later in the election of some Swedish or Norwegian
naturalist (whose name I forget) to the French
Academy of Science instead of Charles Darwin:
the former had described some new kind of fly,
the latter was only a theorizer! The study of
origins in particular was to be frowned upon. In
1863 the Linguistic Society of Paris passed a by-
law that no communications bearing upon the origin of language would be received. In the same
mood, Sir Henry Maine's treatise on "Ancient Law"
was condemned at a leading American university:
it was enough for us to know our own laws; those

of India might interest British students who might have occasion to go there, but not Americans. Such crude notions, utterly hostile to the spirit of science, were unduly favoured fifty years ago by the persistent unwillingness to submit the phenomena of organic nature to the kind of scientific explanation which facts from all quarters were urging upon us.

During the period from 1830 to 1860, the factor in evolution which had hitherto escaped detection was gradually laid hold of and elaborately studied by Charles Darwin. In the nature of his speculations, and the occasion that called them forth, he was a true disciple of Lyell. The work of that great geologist led directly up to Darwinism. As long as it was supposed that each geologic period was separated from the periods before and after it by Titanic convulsions which revolutionized the face of the globe, it was possible for men to acquiesce in the supposition that these convulsions wrought an abrupt and a wholesale destruction of organic life, and that the lost forms were replaced by an equally abrupt and wholesale supernatural creation of new forms at the beginning of each new period. But, as people ceased to believe in the convulsions, such an explanation began to seem improbable, and it was completely discredited by the fact that many kinds of plants and animals have persisted with little or no change during several successive periods, side by side with other kinds in which there have been extensive variation and extinction.

In connection with this a fact of great significance was elicited. Between the fauna and flora of successive periods in the same geographical region there is apt to be a manifest family likeness, indicating that the later are connected with the earlier through the bonds of physical descent. It was a case of this sort that attracted Darwin's attention in 1835. The plants and animals of the Galapagos Islands are either descended, with specific modifications, from those of the mainland of Ecuador, or else there must have been an enormous number of special creations. The case is one which at a glance presents the notion of special creations in an absurd light. But what could have caused the modification? What was wanted was, to be able to point to some agency, similar to agencies now in operation, and therefore intelligible, which could be proved to be capable of making specific changes in plants and animals. Darwin's solution of the problem was so beautiful, it seems now so natural and inevitable, that we may be in danger of forgetting how complicated and

abstruse the problem really was. Starting from
the known experiences of breeders of domestic animals and cultivated plants, and duly considering
the remarkable and sometimes astonishing changes
that are wrought by simple selection, the problem
was to detect among the multifarious phenomena
of organic nature any agency capable of accom-
plishing what man thus accomplishes by selection.
In detecting the agency of natural selection, working perpetually through the preservation of favoured individuals and races in the struggle for
existence, Darwin found the true cause for which
men were waiting. With infinite patience and caution, he applied his method of explanation to one
group of organic phenomena after another, meeting in every quarter with fresh and often unexpected verification. After more than twenty years,
a singular circumstance led him to publish an account of his researches. The same group of facts
had set a younger naturalist to work upon the
same problem, and a similar process of thought
had led to the same solution. Without knowing
what Darwin had done, Alfred Russel Wallace
made the same discovery, and sent from the East
Indies, in 1858, his statement of it to Darwin as
to the man whose judgment upon it he should most
highly prize. This made publication necessary for

Darwin. The vast treasures of theory and example which he had accumulated were given to the world, the notion of special creations was exploded, and the facts of phylogenetic evolution won general acceptance.

Under the influence of this great achievement, men in every department of science began to work in a more philosophical spirit. Naturalists, abandoning the mood of the stamp collectors, saw in every nook and corner some fresh illustration of Darwin's views. One serious obstacle to any general statement of the doctrine of evolution was removed. It was in 1861 that Herbert Spencer began to publish such a general systematic statement. His point of departure was the point reached by Baer in 1829, the change from homogeneity to heterogeneity. The theory of evolution had already received in Spencer's hands a far more complete and philosophical treatment than ever before, when the discovery of natural selection came to supply the one feature which it lacked. Spencer's thought is often more profound than Darwin's, but he would be the first to admit the indispensableness of natural selection to the successful working-out of his own theory.

The work of Spencer is beyond precedent for comprehensiveness and depth. He began by show-

ing that as a generalization of embryology Baer's
law needs important emendations, and he went on to
prove that, as thus rectified, the law of the development of an ovum is the law which covers the evolution of our planetary system, and of life upon the
earth's surface in all its myriad manifestations.
In Spencer's hands, the time-honoured Nebular
Theory propounded by Immanuel Kant in 1755,
the earliest of all scientific theories of evolution,
took on fresh life and meaning ; and at the same
time the theories of Lamarck and Darwin as to
organic evolution were worked up along with his
own profound generalization of the evolution of
mind into one coherent and majestic whole. Mankind have reason to be grateful that the promise
of that daring prospectus which so charmed and
dazzled us in 1860 is at last fulfilled; that after
six-and-thirty years, despite all obstacles and dis-
couragements, the Master's work is virtually done.

Such a synthesis could not have been achieved,
nor even attempted, without the extraordinary
expansion of molecular physics that marked the
first half of the nineteenth century. When Priestley discovered oxygen, the undulatory theory of
light, the basis of all modern physics, had not been
established. It had indeed been propounded as
long ago as 1678 by the illustrious Christian Huy-

ghens, whom we should also remember as the discoverer of Saturn's rings and the inventor of the
pendulum clock. But Huyghens was in advance
of his age, and the overshadowing authority of
Newton, who maintained a rival hypothesis, prevented due attention being paid to the undulatory
theory until the beginning of the present century,
when it was again taken up and demonstrated by
Fresnel and Thomas Young. About the same
time, our fellow countryman, Count Rumford, was
taking the lead in that series of researches which
culminated in the discovery of the mechanical
equivalent of heat by Dr. Joule in 1843. One of
Priestley's earliest books, the one which made him
a doctor of laws and a fellow of the Royal Society,
was a treatise on electricity, published in 1767.
It was a long step from that book to the one in
which the Danish physicist Oersted, in 1820, demonstrated the intimate correlation between electricity and magnetism, thus preparing the way for
Faraday's great discovery of magneto-electric induction in 1831. By the middle of our century
the work in these various departments of physics
had led to the detection of the deepest truth in
science, — the law of correlation and conservation,
which we owe chiefly to Helmholtz, Mayer, and
Grove. It was proved that light and heat, and

the manifestations of force which we group together under the name of electricity, are various modes of undulatory motion transformable one into another; and that, in the operations of nature, energy is never annihilated, but only changed from one form into another. This generalization includes the indestructibility of matter, and thus lies at the bottom of all chemistry and physics and of all science.

Returning to that chemistry with which we started, we may recall two laws that were propounded early in the century, one of which was instantly adopted, while the other had to wait for its day. Dalton's law of definite and multiple proportions has been ever since 1808 the corner stone of chemical science, and the atomic theory by which he sought to explain the law has exercised a profound influence upon all modern speculation. The other law, announced by Avogadro in 1811, that, " under the same conditions of pressure and temperature, equal volumes of all gaseous substances, whether elementary or compound, contain the same number of molecules," was neglected for nearly fifty years, and then, when it was taken up and applied, it remodelled the whole science of chemistry, and threw a flood of light upon the internal constitution of matter. In this direction a new world of speculation is opening up before us, full of wondrous charm. The amazing progress made since Priestley's day may be summed up in a single contrast. In 1781 Cavendish ascertained the bare fact that water is made up of oxygen and hydrogen ; within ninety years from that time Sir William Thomson was able to tell us that " if the drop of water were magnified to the size of the earth, the constituent atoms would be larger than peas, but not so large as billiard balls." Such a statement is confessedly provisional, but, allowing for this, the contrast is no less striking.

Concerning the various and complicated applications of physical science to the arts, by which human life has been so profoundly affected in the present century, a mere catalogue of them would tax our attention to little purpose. As my object in the present sketch is simply to trace the broad outlines of advance in pure science, I pass over these applications, merely observing that the perpetual interaction between theory and practice is such that each new invention is liable to modify the science in which it originated, either by encountering fresh questions or by suggesting new methods, or in both these ways. The work of men like Pasteur and Koch cannot fail to influence biological theory as much as medical practice.

The practical uses of electricity are introducing new features into the whole subject of molecular physics, and in this region, I suspect, we are to look for some of the most striking disclosures of the immediate future.

A word must be said of the historical sciences, which have witnessed as great changes as any others, mainly through the introduction of the comparative method of inquiry. The first two great triumphs of the comparative method were achieved contemporaneously in two fields of inquiry very remote from one another : the one was the work of Cuvier, above mentioned ; the other was the founding of the comparative philology of the Aryan languages by Franz Bopp, in 1816. The work of Bopp exerted as powerful an influence throughout all the historical fields of study as Cuvier exerted in biology. The young men whose minds were receiving their formative impulses between 1825 and 1840, under the various influences of Cuvier and Saint-Hilaire, Lyell, Goethe, Bopp, and other such great leaders, began themselves to come to the foreground as leaders of thought about 1860 : on the one hand, such men as Darwin, Gray, Huxley, and Wallace ; on the other hand, such as Kuhn and Schleicher, Maine, Maurer, Mommsen, Freeman, and Tylor. The point of the comparative method, in whatever field it may be applied, is that it brings before us a great number of objects so nearly alike that we are bound to assume for them an origin and general history in common, while at the same time they present such differences in detail as to suggest that some have advanced further than others in the direction in which all are travelling; some, again, have been abruptly arrested, others perhaps even turned aside from the path. In the attempt to classify such phenomena, whether in the historical or in the physical sciences, the conception of development is presented to the student with irresistible force. In the case of the Aryan languages, no one would think of doubting their descent from a common original : just side by side is the parallel case of one sub-group of the Aryan languages, namely, the seven Romance languages which we know to have been developed out of Latin since the Christian era. In these cases we can study the process of change resulting in forms that are more or less divergent from their originals. In one quarter a form is retained with little modification ; in another it is completely blurred, as the Latin *metipsissimus* becomes *medesimo* in Italian, but *mismo* in Spanish, while in modern French there is nothing left of it but *même*. So in San-

skrit and in Lithuanian we find a most ingenious
and elaborate system of conjugation and declen-
sion, which in such languages as **Greek** and **Latin**
is more or less curtailed and altered, and which in
English is almost completely lost. Yet in Old
English there are quite enough vestiges of the system to enable us to identify it with the Lithuanian and Sanskrit.

So the student who applies the comparative
method to the study of human customs and institutions is continually finding usages, beliefs, or
laws existing in one part of the world that have
long since ceased to exist in another part; yet
where they have ceased to exist they have often
left unmistakable traces of their former existence.
In Australasia we find types of savagery ignorant
of the bow and arrow; in aboriginal North America, a type of barbarism familiar with the art of
pottery, but ignorant of domestic animals or of
the use of metals; among the earliest Romans, a
higher type of barbarism, familiar with iron and
cattle, but ignorant of the alphabet. Along with
such gradations in material culture we find associated gradations in ideas, in social structure,
and in deep-seated customs. Thus, some kind of
fetishism is apt to prevail in the lower stages of
barbarism, and some form of polytheism in the

higher stages. The units of composition in savage and barbarous societies are always the clan, the phratry, and the tribe. In the lower stages of barbarism we see such confederacies as those of the Iroquois; in the highest stage, at the dawn of civilization, we begin to find nations imperfectly formed by conquest without incorporation, like aboriginal Perú or ancient Assyria. In the lower stages we see captives tortured to death, then at a later stage sacrificed to the tutelar deities, then later on enslaved and compelled to till the soil. Through the earlier stages of culture, as in Australasia and aboriginal America, we find the marriage tie so loose and paternity so uncertain that kinship is reckoned only through the mother; but in the highest stage of barbarism, as among the earliest Greeks, Romans, and Jews, the more definite patriarchal family is developed, and kinship begins to be reckoned through the father. It is only after that stage is reached that inheritance of property becomes fully developed, with the substitution of individual ownership for clan ownership, and so on to the development of testamentary succession, individual responsibility for delict and crime, and the substitution of contract for status. In all such instances — and countless others might be cited — we see the marks of an intelligible progression, a line of development which human ideas and institutions have followed. But in the most advanced societies we find numerous traces of such states of things as now exist only among savage or barbarous societies. Our own ancestors were once polytheists, with plenty of traces of fetishism. They were organized in clans, phratries, and tribes. There was a time when they used none but stone tools and weapons; when there was no private property in land, and no political structure higher than the tribe. Among the forefathers of the present civilized inhabitants of Europe are unmistakable traces of human sacrifices, and of the reckoning of kinship through the mother only. When we have come to survey large groups of facts of this sort, the conclusion is irresistibly driven home to us that the more advanced societies have gone through various stages now represented here and there by less advanced societies; that there is a general path of social development, along which, owing to special circumstances, some peoples have advanced a great way, some a less way, some but a very little way; and that by studying existing savages and barbarians we get a valuable clue to the interpretation of prehistoric times. All these things are to-day commonplaces among students of history and archæology; sixty years ago they

would have been scouted as idle vagaries. It is
the introduction of such methods of study that is
making history scientific. It is enabling us to digest the huge masses of facts that are daily poured
in upon us by decipherers of the past, — monu-
ments, inscriptions, pottery, weapons, ethnological
reports, and all that sort of thing, — and to make
all contribute toward a coherent theory of the
career of mankind upon the earth.

In the course of the foregoing survey one fact
stands out with especial prominence : it appears
that about half a century ago the foremost minds
of the world, with whatever group of phenomena
they were occupied, had fallen, and were more and
more falling, into a habit of regarding things, not
as having originated in the shape in which we now
find them, but as having been slowly metamor-
phosed from some other shape through the agency
of forces similar in nature to forces now at work.
Whether planets, or mountains, or mollusks, or
subjunctive moods, or tribal confederacies were the
things studied, the scholars who studied them most
deeply and most fruitfully were those who studied
them as phases in a process of development. The
work of such scholars has formed the strong current of thought in our time, while the work of
those who did not catch these new methods has

been dropped by the way and forgotten; and as we look back to Newton's time we can see that ever since then the drift of scientific thought has been setting in this direction, and with increasing steadiness and force.

Now, what does all this drift of scientific opinion during more than two centuries mean? It can, of course, have but one meaning. It means that the world *is* in a process of development, and that gradually, as advancing knowledge has enabled us to take a sufficiently wide view of the world, we have come to see that it is so. The old statical conception of a world created all at once in its present shape was the result of very narrow experience; it was entertained when we knew only an extremely small segment of the world. Now that our experience has widened, it is outgrown and set aside forever; it is replaced by the dynamical conception of a world in a perpetual process of evolution from one state into another state. This dynamical conception has come to stay with us. Our theories as to what the process of evolution is may be more or less wrong and are confessedly tentative, as scientific theories should be. But the dynamical conception, which is not the work of any one man, be he Darwin or Spencer or any one else, but the result of the cumulative experience of the last two centuries, — this is a permanent acquisition. We can no more revert to the statical conception than we can turn back the sun in his course. Whatever else the philosophy of future generations may be, it must be some kind of a philosophy of evolution.

Such is the scientific conquest achieved by the nineteenth century, a marvellous story without any parallel in the history of human achievement. The swiftness of the advance has been due partly to the removal of the ancient legal and social trammels that beset free thinking in every conceivable direction. It is largely due also to the use of correct methods of research. The waste of intellectual effort has been less than in former ages. The substitution of Lavoisier's balance for Stahl's *a priori* reasoning is one among countless instances of this. Sound scientific method is a slow acquisition of the human mind, and for its more rapid introduction, in Priestley's time and since, we have largely to thank the example set by those giants of a former age, Galileo and Kepler, Descartes and Newton.

The lessons that might be derived from our story are many. But one that we may especially emphasize is the dignity of Man whose persistent seeking for truth is rewarded by such fruits. We may be sure that the creature whose intelligence measures

the pulsations of molecules and unravels the secret
of the whirling nebula is no creature of a day, but
the child of the universe, the heir of all the ages,
in whose making and perfecting is to be found the
consummation of God's creative work.

May, 1896.

II

THE DOCTRINE OF EVOLUTION: ITS SCOPE AND PURPORT[1]

IT was not strange that among the younger men whose opinions were moulded between 1830 and 1840 there should have been one of organizing genius, with a mind inexhaustibly fertile in suggestions, who should undertake to elaborate a general doctrine of evolution, to embrace in one grand coherent system of generalizations all the minor generalizations which workers in different departments of science were establishing. It is this prodigious work of construction that we owe to Herbert Spencer. He is the originator and author of what we know to-day as the doctrine of evolution, the doctrine which undertakes to formulate and put into scientific shape the conception of evolution toward which scientific investigation had so long been tending. In the mind of the general public there seems to be dire confusion with regard to Mr. Spencer

[1] Part of an address before the Brooklyn Ethical Association, May 31, 1891.

and his relations to evolution and to Darwinism.
Sometimes, I believe, he is even supposed to be
chiefly a follower and expounder of Mr. Darwin!
No doubt this is because so many people mix up
Darwinism with the doctrine of evolution, and have
but the vaguest and haziest notions as to what it is
all about. As I explained above, Mr. Darwin's
great work was the discovery of natural selection,
and the demonstration of its agency in effecting
specific changes in plants and animals; and in
that work he was completely original. But plants
and animals are only a part of the universe, though
an important part, and with regard to universal
evolution or any universal formula for evolution
Darwinism had nothing to say. Such problems
were beyond its scope.

The discovery of a universal formula for evolution, and the application of this formula to many
diverse groups of phenomena, have been the great
work of Mr. Spencer, and in this he has had no
predecessor. His wealth of originality is immense,
and it is unquestionable. But as the most original
thinker must take his start from the general stock
of ideas accumulated at his epoch, and more often
than not begins by following a clue given him by
somebody else, so it was with Mr. Spencer when,
about forty years ago, he was working out his doc-

trine of evolution. The clue was not given by
Mr. Darwin. Darwinism was not yet born. Mr.
Spencer's theory was worked out in all its parts,
and many parts of it had been expounded in various published volumes and essays before the publication of the " Origin of Species."

The clue which Mr. Spencer followed was given
him by the great embryologist, Karl Ernst von
Baer, and an adumbration of it may perhaps
be traced back through Kaspar Friedrich Wolf
to Linnæus. Hints of it may be found, too, in
Goethe and in Schelling. The advance from simplicity to complexity in the development of an
egg is too obvious to be overlooked by any one,
and was remarked upon, I believe, by Harvey ;
but the analysis of what that advance consists
in was a wonderfully suggestive piece of work.
Baer's great book was published in 1829, just at
the time when so many stimulating ideas were
being enunciated, and its significant title was
Entwickelungsgeschichte, or " History of Evolution." It was well known that, so far as the
senses can tell us, one ovum is indistinguishable
from another, whether it be that of a man, a fish,
or a parrot. The ovum is a structureless bit of
organic matter, and, in acquiring structure along
with its growth in volume and mass, it proceeds

through a series of differentiations, and the result
is a change from homogeneity to heterogeneity.
Such was Baer's conclusion, to which scanty justice is done by such a brief statement. As all
know, his work marked an epoch in the study
of embryology; for to mark the successive differentiations in the embryos of a thousand animals
was to write a thousand life histories upon correct
principles.

Here it was that Mr. Spencer started. As a
young man, he was chiefly interested in the study
of political government and in history so far as it
helps the study of politics. A philosophical student
of such subjects must naturally seek for a theory
of evolution. If I may cite my own experience, it
was largely the absorbing and overmastering passion for the study of history that first led me to
study evolution in order to obtain a correct method.
When one has frequent occasion to refer to the
political and social *progress* of the human race,
one likes to know what one is talking about. Mr.
Spencer needed a theory of progress. He could
see that the civilized part of mankind has undergone some change from a bestial, unsocial, perpetually fighting stage of savagery into a partially
peaceful and comparatively humane and social
stage, and that we may reasonably hope that the

change in this direction will go on. He could see, too, that along with this change there has been a building-up of tribes into nations, a division of labour, a differentiation of governmental functions, a series of changes in the relations of the individual to the community. To see so much as this is to whet one's craving for enlarged resources wherewith to study human progress. Mr. Spencer had a wide, accurate, and often profound acquaintance with botany, zoölogy, and allied studies. The question naturally occurred to him, Where do we find the process of development most completely exemplified from beginning to end, so that we can follow and exhaustively describe its consecutive phases? Obviously in the development of the ovum. There, and only there, do we get the whole process under our eyes from the first segmentation of the yolk to the death of the matured individual. In other groups of phenomena we can only see a small part of what is going on; they are too vast for us, as in astronomy, or too complicated, as in sociology. Elsewhere our evidences of development are more or less piecemeal and scattered, but in embryology we do get, at any rate, a connected story.

So Mr. Spencer took up Baer's problem, and carried the solution of it much further than the great Esthonian naturalist. He showed that in the development of the ovum the change from homogeneity to heterogeneity is accompanied by a change from indefiniteness to definiteness; there are segregations of similarly differentiated units resulting in the formation of definite organs. He further showed that there is a parallel and equally important change from incoherence to coherence; along with the division of labour among the units there is an organization of labour: at first, among the homogeneous units there is no subordination, — to subtract one would not alter the general aspect; but at last, among the heterogeneous organs there is such subordination and interdependence that to subtract any one is liable to undo the whole process and destroy the organism. In other words, integration is as much a feature of development as differentiation; the change is not simply from a structureless whole into parts, but it is from a structureless whole into an organized whole with a consensus of different functions, and that is what we call an organism. So while Baer said that the evolution of the chick is a change from homogeneity to heterogeneity through successive differentiations, Mr. Spencer said that the evolution of the chick is a continuous change from indefinite incoherent homogeneity to definite coherent heterogeneity through successive differentiations and integrations.

But Mr. Spencer had now done something more than describe exhaustively the evolution of an individual organism. He had got a standard of high and low degrees of organization; and the next thing in order was to apply this standard to the whole hierarchy of animals and plants according to their classified relationships and their succession in geological time. This was done with most brilliant success. From the earliest records in the rocks, the general advance in types of organization has been an advance in definiteness, coherence, and heterogeneity. The method of evolution in the life history of the animal and vegetal kingdoms has been like the method of evolution in the life history of the individual.

To go into the inorganic world with such a formula might seem rash. But as the growth of organization is essentially a particular kind of redistribution of matter and motion, and as redistribution of matter and motion is going on universally in the inorganic world, it is interesting to inquire whether, in such simple approaches toward organization as we find, there is any approach toward the characteristics of organic evolution as above described. It was easy for Mr. Spencer to show that the change from a nebula into a planetary system conforms to the definition of evolution in a way that is most striking and suggestive. But in studying the inorganic world Mr. Spencer was led to modify his formula in a way that vastly increased its scope. He came to see that the primary feature of evolution is an integration of matter and concomitant dissipation of motion. According to circumstances, this process may or not be attended with extensive internal rearrangements and development of organization. The continuous internal rearrangement implied in the development of organization is possible only where there is a medium degree of mobility among the particles, a plasticity such as is secured only by those peculiar chemical combinations which make up what we call organic matter. In the inorganic world, where there is an approach to organization there is an adumbration of the law as realized in the organic world. But in the former, what strikes us most is the concentration of the mass with the retention of but little internal mobility; in the latter, what strikes us most is the wonderful complication of the transformations wrought by the immense amount of internal mobility retained. These transformations are to us the mark, the distinguishing feature, of life.

Having thus got the nature of the differences between the organic and inorganic worlds into a series of suggestive formulas, the next thing to be done was to inquire into the applicability of the law of evolution to the higher manifestations of vital activity, — in other words, to psychical and social life.   Here it was easy to point out analogies between the development of society and the development of an organism.   Between a savage state of society and a civilized state, it is easy to see the contrasts in complexity of life, in division of labour, in interdependence and coherence of operations and of interests.   The difference resembles that between a vertebrate animal and a worm.

Such analogies are instructive, because at the bottom of the phenomena there is a certain amount of real identity.   But Mr. Spencer did not stop with analogies; he pursued his problem into much deeper regions.   There is one manifest distinction between a society and an organism.   In the organism, the conscious life, the psychical life, is not in the parts, but in the whole; but in a society, there is no such thing as corporate consciousness: the psychical life is all in the individual men and women. The highest development of this psychical life is the end for which the world exists.   The object of social life is the highest spiritual welfare of the individual members of society. The individual human soul thus comes to be as much the centre of the Spencerian world as it was the centre of the world of mediæval theology; and the history of the evolution of conscious intelligence becomes a theme of surpassing interest.

This is the part of his subject which Mr. Spencer has handled in the most masterly manner. Nothing in the literature of psychology is more remarkable than the long-sustained analysis in which he starts with complicated acts of quantitative reasoning and resolves them into their elementary processes, and then goes on to simpler acts of judgment and perception, and then down to sensation, and so on resolving and resolving, until he gets down to the simple homogeneous psychical shocks or pulses in the manifold compounding and recompounding of which all mental action consists. Then, starting afresh from that conception of life as the continuous adjustment of inner relations within the organism to outer relations in the environment, — a conception of which he made such brilliant use in his "Principles of Biology," — he shows how the psychical life gradually becomes specialized in certain classes of adjustments or correspondences, and how the development of psychical life consists in a progressive differentiation and integration of such correspondences. Intellectual life is shown to have arisen by slow gradations, and the special interpretations of reflex action, instinct, memory, reason, emotion, and will are such as to make the " Principles of Psychology " indubitably the most suggestive book upon mental phenomena that was ever written.

Toward the end of the first edition of the " Origin of Species," published in 1859, Mr. Darwin looked forward to a distant future when the conception of gradual development might be applied to the phenomena of intelligence. But the first edition of the " Principles of Psychology," in which this was so successfully done, had already been published four years before, — in 1855, — so that Mr. Darwin in later editions was obliged to modify his statement, and confess that, instead of looking so far forward, he had better have looked about him. I remember hearing Mr. Darwin laugh merrily over this at his own expense.

This extension of the doctrine of evolution to psychical phenomena was what made it a universal doctrine, an account of the way in which the world, as we know it, has been evolved. There is no subject, great or small, that has not come to be affected by the doctrine, and, whether men realize it or not, there is no nook or corner in speculative science where they can get away from the sweep of Mr. Spencer's thought.

This extension of the doctrine to psychical phenomena is by many people misunderstood.   The
" Principles of Psychology " is a marvel of straightforward and lucid statement ; but, from its immense
reach and from the abstruseness of the subject, it
is not easy reading.   It requires a sustained attention such as few people can command, except on
subjects with which they are already familiar.
Hence few people read it in comparison with the
number who have somehow got it into their heads
that Mr. Spencer tries to explain mind as evolved
out of matter, and is therefore a materialist.   How
many worthy critics have been heard to object to
the doctrine of evolution that you cannot deduce
mind from the primeval nebula, unless the germs
of mind were present already !   But that is just
what Mr. Spencer says himself.   I have heard
him say it more than once, and his books contain
many passages of equivalent import.[1]   He never
misses an opportunity for attacking the doctrine
that mind can be explained as evolved from matter.   But, in spite of this, a great many people suppose that the gradual evolution of mind *must* mean
its evolution out of matter, and are deaf to arguments of which they do not perceive the bearing.

[1] See, for example, *Principles of Psychology*, second edition,
1870–72, vol. ii. pp. 145–162.

Hence Mr. Spencer is so commonly accredited with the doctrine which he so earnestly repudiates.

But there is another reason why people are apt to suppose the doctrine of evolution to be materialistic in its implications. There are able writers who have done good service in illustrating portions of the general doctrine, and are at the same time avowed materialists. One may be a materialist, whatever his scientific theory of things; and to such a person the materialism naturally seems to be a logical consequence from the scientific theory. We have received this evening a communication from Professor Ernst Haeckel, of Jena, in which he lays down five theses regarding the doctrine of evolution : —

1. " The general doctrine appears to be already unassailably founded;

2. " Thereby every supernatural creation is completely excluded ; .

3. " Transformism and the theory of descent are inseparable constituent parts of the doctrine of evolution ;

4. " The necessary consequence of this last conclusion is the descent of man from a series of vertebrates."

So far, very good ; we are within the limits of scientific competence, where Professor Haeckel is

strong. But now, in his fifth thesis, he enters the
region of metaphysics, — the transcendental region,
which science has no competent methods of exploring, — and commits himself to a dogmatic assertion :

5. "The beliefs in an 'immortal soul' and in
'a personal God' are therewith" (*i. e.*, with the
four preceding statements) "completely ununitable
(*völlig unvereinbar*)."

Now, if Professor Haeckel had contented himself
with asserting that these two beliefs are not susceptible of scientific demonstration ; if he had simply
said that they are beliefs concerning which a scientific man, in his scientific capacity, ought to refrain
from making assertions, because Science knows nothing whatever about the subject, he would have
occupied an impregnable position. His fifth thesis would have been as indisputable as his first four.
But Professor Haeckel does not stop here. He declares virtually that if an evolutionist is found
entertaining the beliefs in a personal God and an immortal soul, nevertheless these beliefs are not philosophically reconcilable with his scientific theory of
things, but are mere remnants of an old-fashioned
superstition from which he has not succeeded in
freeing himself.

Here one must pause to inquire what Professor
Haeckel means by "a personal God." If he refers

to the Latin conception of a God remote from the
world of phenomena, and manifested only through
occasional interference, — the conception that has
until lately prevailed in the Western world since
the time of St. Augustine, — then we may agree
with him; the practical effect of the doctrine of
evolution is to abolish such a conception. But
with regard to the Greek conception entertained
by St. Athanasius; the conception of God as im-
manent in the world of phenomena and manifested
in every throb of its mighty rhythmical life; the
deity that Richard Hooker, prince of English
churchmen, had in mind when he wrote of Natural
Law that "her seat is the bosom of God, and her
voice the harmony of the world," — with regard
to this conception the practical effect of the doctrine of evolution is not to abolish, but to strengthen
and confirm it. For, into whatever province of
Nature we carry our researches, the more deeply
we penetrate into its laws and methods of action,
the more clearly do we see that all provinces of
Nature are parts of an organic whole animated by
a single principle of life that is infinite and eternal. I have no doubt Professor Haeckel would
not only admit this, but would scout any other view
as inconsistent with the monism which he professes.
But he would say that this infinite and eternal

principle of life is not psychical, and therefore cannot be called in any sense " a personal God." In an ultimate analysis, I suspect Professor Haeckel's ubiquitous monistic principle would turn out to be neither more nor less than Dr. Büchner's mechanical force (*Kraft*). On the other hand, I have sought to show — in my little book " The Idea of God " — that the Infinite and Eternal Power that animates the universe must be psychical in its nature, that any attempt to reduce it to mechanical force must end in absurdity, and that the only kind of monism which will stand the test of an ultimate analysis is monotheism. While in the chapter on Anthropomorphic Theism, in my " Cosmic Philosophy," I have taken great pains to point out the difficulties in which (as finite thinkers) we are involved when we try to conceive the Infinite and Eternal Power as psychical in his nature, I have in the chapter on Matter and Spirit, in that same book, taken equal pains to show that we are logically compelled thus to conceive Him.

One's attitude toward such problems is likely to be determined by one's fundamental conception of psychical life. To a materialist the ultimate power is mechanical force, and psychical life is nothing but the temporary and local result of fleeting collocations of material elements in the shape of nervous systems.   Into the endless circuit of transformations of molecular motion, says the materialist, there enter certain phases which we call feelings and thoughts; they are part of the circuit; they arise out of motions of material molecules, and disappear by being retransformed into such motions: hence, with the death of the organism in which such motions have been temporarily gathered into a kind of unity, all psychical activity and all personality are *ipso facto* abolished.   Such is the materialistic doctrine, and such, I presume, is what Professor Haeckel has in mind when he asserts that the belief in an immortal soul is incompatible with the doctrine of evolution.   The theory commonly called that of the correlation of forces, and which might equally well or better be called the theory of the metamorphosis of motions, is indispensable to the doctrine of evolution.   But for the theory that light, heat, electricity, and nerve-action are different modes of undulatory motion transformable one into another, and that similar modes of motion are liberated by the chemical processes going on within the animal or vegetal organism, Mr. Spencer's work could never have been done.   That theory of correlation and transformation is now generally accepted, and is often appealed to by materialists.   A century ago Cabanis said that the

brain secretes thought as the liver secretes bile.
If he were alive to-day, he would doubtless smile
at this old form of expression as crude, and would
adopt a more subtle phrase; he would say that
" thought is transformed motion."

Against this interpretation I have maintained
that the theory of correlation not only fails to support it, but actually overthrows it.   The arguments
may be found in the chapter on **Matter** and Spirit,
in my " Cosmic Philosophy," published in 1874, and
in the essay entitled " A Crumb for the Modern
Symposium," written in 1877, and reprinted in
" Darwinism and Other Essays." [1]   Their purport is,
that in tracing the correlation of motions into the organism through the nervous system and out again,
we are bound to get an account of each step in
terms of motion.   Unless we can show that every
unit of motion that disappears is transformed into
an exact quantitative equivalent, our theory of correlation breaks down ; but when we have shown this
we shall have given a complete account of the
whole affair without taking any heed whatever of
thought, feeling, or consciousness.   In other words,
these psychical activities do not enter into the circuit, but stand outside of it, as a segment of a
circle may stand outside a portion of an entire cir-

---

[1] See also *Excursions of an Evolutionist*, 1883, pp. 274–282.

cumference with which it is concentric. Motion is never transformed into thought, but only into some other form of measurable (in fact, or at any rate in theory, measurable) motion that takes place in nerve-threads and ganglia. *It is not the thought, but the nerve-action that accompanies the thought, that is really " transformed motion."* I say that if we are going to verify the theory of correlation, it must be done (actually or theoretically) by measurement; quantitative equivalence must be proved at every step; and hence we must not change our unit of measurement; from first to last it must be a unit of motion: if we change it for a moment, our theory of correlation that moment collapses. I say, therefore, that the theory of correlation and equivalence of forces lends no support whatever to materialism. On the contrary, its manifest implication is that psychical life cannot be a mere product of temporary collocations of matter.

The argument here set forth is my own. When I first used it, I had never met with it anywhere in books or conversation. Whether it has since been employed by other writers I do not know, for during the past fifteen years I have read very few books on such subjects. At all events, it is an argument for which I am ready to bear the full

responsibility.   Some doubt has recently been expressed whether Mr. Spencer would admit the force
of this argument.   It has been urged by Mr. S. H.
Wilder, in two able papers published in the " New
York Daily Tribune," June 13 and July 4, 1890,
that the use of this argument marks a radical
divergence on my part from Mr. Spencer's own
position.

It is true that in several passages of " First Principles" there are statements which either imply or
distinctly assert that motion can be transformed
into feeling and thought, — *e. g. :* " Those modes of
the Unknowable which we call heat, light, chemical affinity, etc., are alike transformable into each
other, and into those modes of the Unknowable
which we distinguish as sensation, emotion, thought;
these, in their turns, being directly or indirectly re-
transformable into the original shapes ; "[1] and again,
it is said " to be a necessary deduction from the
law of correlation that what exists in consciousness
under the form of feeling is transformable into
an equivalent of mechanical motion," etc.[2]   Now,
if this, as literally interpreted, be Mr. Spencer's
deliberate opinion, I entirely dissent from it.   To
speak of quantitative equivalence between a unit

[1] *First Principles*, second edition, 1867, p. 217.
[2] *Id.* p. 558.

of feeling and a unit of motion seems to me to be talking nonsense, — to be combining terms which severally possess a meaning into a phrase which has no meaning. I am therefore inclined to think that the above sentences, literally interpreted, do not really convey Mr. Spencer's opinion. They appear manifestly inconsistent, moreover, with other passages in which he has taken much more pains to explain his position.[1] In the sentence from page 558 of " First Principles," Mr. Spencer appears to me to mean that the nerve-action, which is the objective concomitant of what is subjectively known as feeling, is transformable into an equivalent of mechanical motion. When he wrote that sentence perhaps he had not shaped the case quite so distinctly in his own mind as he had a few years later, when he made the more elaborate statements in the second edition of the Psychology. Though in these more elaborate statements he does not assert the doctrine I have here maintained, yet they seem consistent with it. When I was finishing the chapter on Matter and Spirit, in my room in London, one afternoon in February, 1874, Mr. Spencer came in, and I read to him nearly the whole chapter, including my

[1] See, e. g., *Principles of Psychology*, second edition, vol. i. pp. 158-161, 616-627.

argument from correlation above mentioned. He expressed warm approval of the chapter, without making any specific qualifications. In the course of the chapter I had occasion to quote a passage from the Psychology,[1] in which Mr. Spencer twice inadvertently used the phrase "nervous shock" where he meant "psychical shock." As his object was to keep the psychical phenomena and their cerebral concomitants distinct in his argument, this colloquial use of the word "nervous" was liable to puzzle the reader, and give querulous critics a chance to charge Mr. Spencer with the materialistic implications which it was his express purpose to avoid. Accordingly, in my quotation I changed the word "nervous" to "psychical," using brackets and explaining my reasons. On showing all this to Mr. Spencer, he desired me to add in a footnote that he thoroughly approved the emendation.

I mention this incident because our common, every-day speech abounds in expressions that have a materialistic flavour; and sometimes in serious writing an author's sheer intentness upon his main argument may lead him to overlook some familiar form of expression which, when thrown into a precise and formal context, will strike the reader in a

---

[1] Vol. i. p. 158. Cf. my *Cosmic Philosophy*, vol. ii. p. 444.

very different way from what the author intended. I am inclined to explain in this way the passages in " First Principles " which are perhaps chiefly responsible for the charge of materialism that has so often and so wrongly been brought up against the doctrine of evolution.

As regards the theological implications of the doctrine of evolution, I have never undertaken to speak for Mr. Spencer; on such transcendental subjects it is quite enough if one speaks for one's self. It is told of Diogenes that, on listening one day to a sophistical argument against the possibility of motion, he grimly got up out of his tub and walked across the street. Whether his adversaries were convinced or not, we are not told. Probably not; it is but seldom that adversaries are convinced. So, when Professor Haeckel declares that belief in a " personal God " and an " immortal soul " is incompatible with acceptance of the doctrine of evolution, I can only say, for myself — however much or little the personal experience may be worth — I find that the beliefs in the psychical nature of God and in the immortality of the human soul seem to harmonize infinitely better with my general system of cosmic philosophy than the negation of these beliefs. If Professor Haeckel, or any other writer, prefers a materialistic

interpretation, very well. I neither quarrel with
him nor seek to convert him ; but I do not agree
with him. I do not pretend that my opinion on
these matters is susceptible of scientific demonstration. Neither is his. I say, then, that his fifth
thesis has no business in a series of scientific generalizations about the doctrine of evolution.

Far beyond the limits of what scientific methods,
based upon our brief terrestrial experience, can de-
monstrate, there lies on every side a region with
regard to which Science can only suggest questions.
As Goethe so profoundly says : —

> " Willst du ins Unendliche streiten,
> Geh' nur im Endlichen nach allen Seiten." [1]

It is of surpassing interest that the particular generalization which has been extended into a universal formula of evolution should have been the
generalization of the development of an ovum.
In enlarging the sphere of life in such wise as to
make the whole universe seem actuated by a single
principle of life, we are introduced to regions of
sublime speculation. The doctrine of evolution,
which affects our thought about all things, brings
before us with vividness the conception of an ever
present God, — not an absentee God who once

[1] "If thou wouldst press into the infinite, go but to all parts
of the finite."

manufactured a cosmic machine capable of running itself, except for a little jog or poke here and there in the shape of a special providence. The doctrine of evolution destroys the conception of the world as a machine. It makes God our constant refuge and support, and Nature his true revelation; and when all its religious implications shall have been set forth, it will be seen to be the most potent ally that Christianity has ever had in elevating mankind.

March, 1890.

III

EDWARD LIVINGSTON YOUMANS [1]

IN one of the most beautiful of all the shining pages of his " History of the Spanish Conquest in America," Sir Arthur Helps describes the way in which, through " some fitness of the season, whether in great scientific discoveries or in the breaking into light of some great moral cause, the same processes are going on in many minds, and it seems as if they communicated with each other invisibly. We may imagine that all good powers aid the ' new light,' and brave and wise thoughts about it float aloft in the atmosphere of thought as downy seeds are borne over the fruitful face of the earth." [2] The thinker who elaborates a new system of philosophy, deeper and more comprehensive than any yet known to mankind, though he may work in solitude, nevertheless does not work alone. The very fact which makes his great scheme of thought a success, and not a failure, is the fact

[1] An address before the Brooklyn Ethical Association, March 23, 1890.

[2] Vol. iii. p. 113.

that it puts into definite and coherent shape the ideas which many people are ,more or less vaguely and loosely entertaining, and that it carries to a grand and triumphant conclusion processes of reasoning in which many persons have already begun taking the earlier steps. This community in mental trend between the immortal discoverer and many of the brightest contemporary minds, far from diminishing the originality of his work, constitutes the feature of it which makes it a permanent acquisition for mankind, and distinguishes it from the eccentric philosophies which now and then come up to startle the world for a while, and are presently discarded and forgotten. The history of modern physics — as in the case of the correlation of forces and the undulatory theory of light — furnishes us with many instances of wise thoughts floating like downy seeds in the atmosphere until the moment has come for them to take root. And so it has been with the greatest achievement of modern thinking, — the doctrine of evolution. Students and investigators in all departments, alike in the physical and in the historical sciences, were fairly driven by the nature of the phenomena before them into some hypothesis, more or less vague, of gradual and orderly change or development. The world was ready and waiting for Herbert Spencer's mighty work when it came, and it was for that reason that it was so quickly triumphant over the old order of thought. The victory has been so thorough, swift, and decisive that it will take another generation to narrate the story of it so as to do it full justice. Meanwhile, people's minds are apt to be somewhat dazed with the rapidity and wholesale character of the change; and nothing is more common than to see them adopting Mr. Spencer's ideas without recognizing them as his or knowing whence they got them. As fast as Mr. Spencer could set forth his generalizations they were taken hold of here and there by special workers, each in his own department, and utilized therein. His general system was at once seized, assimilated, and set forth with new illustrations by serious thinkers who were already groping in the regions of abstruse thought which the master's vision pierced so clearly. And thus the doctrine of evolution has come to be inseparably interfused with the whole mass of thinking in our day and generation. I do not mean to imply that people commonly entertain very clear ideas about it, for clear ideas are not altogether common. I suspect that a good many people would hesitate if asked to state exactly what Newton's law of gravitation is.

Among the men in America whose minds, between thirty and forty years ago, were feeling their way toward some such unified conception of nature as Mr. Spencer was about to set forth in all its dazzling glory, — among the men who were thus prepared to grasp the doctrine of evolution at once and expound it with fresh illustrations, — the first in the field was the man to whose memory we have met here this evening to pay a brief word of tribute. It is but a little while since that noble face was here with us, and the tones of that kindly voice were fraught with good cheer for us. To most of you, I presume, the man Edward Livingston Youmans is still a familiar presence. There must be many here this evening who listened to the tidings of his death three years ago with a sense of personal bereavement. No one who knew him is likely ever to forget him. But for those who remember distinctly the man it may not be superfluous to recount the principal incidents of his life and work. It is desirable that the story should be set forth concisely, so as to be remembered ; for the work was like the man, unselfish and unobtrusive, and in the hurry and complication of modern life such work is liable to be lost from sight, so that people profit by it without knowing that it was ever done. So genuinely modest, so utterly destitute of self-

regarding impulses, was our friend, that I believe
it would be quite like him to chide us for thus
drawing public attention to him, as he would
think, with too much emphasis. But such mild
reproof it is right that we should disregard; for
the memory of a life so beautiful and useful is a
precious possession of which mankind ought not to
be deprived.

Edward Livingston Youmans was born in the
town of Coeymans, Albany County, N. Y., on the
3d of June, 1821. From his father and mother,
both of whom survived him, he inherited strong
traits of character as well as an immense fund
of vital energy, such that the failure of health a
few years ago seemed (to me, at least) surprising. His father, Vincent Youmans, was a man of
independent character, strong convictions, and perfect moral courage, with a quick and ready tongue,
in the use of which earnestness and frankness perhaps sometimes prevailed over prudence. The
mother, Catherine Scofield, was notable for balance of judgment, prudence, and tact. The
mother's grandfather was Irish; and while I very
much doubt the soundness of the generalizations
we are so prone to make about race characteristics,
I cannot but feel that for the impulsive — one had
almost said explosive — warmth of sympathy, the

enchanting grace and vivacity of manner, in Edward Youmans, this strain of Irish blood may have been to some extent accountable. Both father and mother belonged to the old Puritan stock of New England, and the father's ancestry was doubtless purely English. Nothing could be more honourably or characteristically English than the name. In the old feudal society, the *yeoman*, like the *franklin*, was the small freeholder, owning a modest estate, yet holding it by no servile tenure; a man of the common people, yet no churl; a member of the state who " knew his rights, and knowing dared maintain." Few indeed were the nooks and corners outside of merry England where such men flourished as the yeomen and franklins who founded democratic New England. It has often been remarked how the most illustrious of Franklins exemplified the typical virtues of his class. There was much that was similar in the temperament and disposition of Edward Youmans, — the sagacity and penetration, the broad common sense, the earnest purpose veiled but not hidden by the blithe humour, the devotion to ends of wide practical value, the habit of making in the best sense the most out of life.

When Edward was but six months old, his parents moved to Greenfield, near Saratoga Springs.

With a comfortable house and three acres of land,
his father kept a wagon shop and smithy. In
those days, while it was hard work to wring a sub-
sistence out of the soil or to prosper upon any of
the vocations which rural life permitted, there was
doubtless more independence of character and real
thriftiness than in our time, when cities and tariffs
have so sapped the strength of the farming country. In the family of Vincent Youmans, though
rigid economy was practised, books were reckoned
to a certain extent among the necessaries of life,
and the house was one in which neighbours were
fond of gathering to discuss questions of politics or
theology, social reform or improvements in agriculture. On all such questions Vincent Youmans
was apt to have ideas of his own ; he talked with
enthusiasm, and was also ready to listen ; and he
evidently supplied an intellectual stimulus to the
whole community. For a boy of bright and inquisitive mind, listening to such talk is no mean source
of education. It often goes much further than
the reading of books. From an early age Edward
Youmans seems to have appropriated all such
means of instruction. He had that insatiable
thirst for knowledge which is one of God's best
gifts to man ; for he who is born with this appetite
must needs be grievously ill made in other respects

if it does not constrain him to lead a happy and useful life.

After ten years at Greenfield the family moved to a farm at Milton, some two miles distant. Until his sixteenth year Edward helped his father at farm work in the summer, and attended the district school in winter. It was his good fortune at that time to fall into the hands of a teacher who had a genius for teaching, — a man who in those days of rote-learning did not care to have things learned by heart, but sought to stimulate the thinking powers of his pupils, and who in that age of canes and ferules never found it necessary to use such means of discipline, because the fear of displeasing him was of itself all-sufficient. Experience of the methods of such a man was enough to sharpen one's disgust for the excessive mechanism, the rigid and stupid manner of teaching, which characterize the ordinary school. In after years Youmans used to say that "Uncle Good" — as this admirable pedagogue was called — first taught him what his mind was for. Through intercourse and training of this sort he learned to doubt, to test the soundness of opinions, to make original inquiries, and to find and follow clues.

But even the best of teachers can effect but little unless he finds a mind ready of itself to take

the initiative. It is doubtful if men of eminent
ability are ever made so by schooling. The school
offers opportunities, but in such men the tendency
to the initiative is so strong that if opportunities
are not offered they will somehow contrive to
create them. When Edward Youmans was about
thirteen years old he persuaded his father to buy
him a copy of Comstock's Natural Philosophy.
This book he studied at home by himself, and repeated many of the experiments with apparatus of
his own contriving. When he made a centrifugal
water wheel, and explained to the men and boys
of the neighbourhood the principle of its revolution
in a direction opposite to that of the stream which
moved it, we may regard it as his earliest attempt
at giving scientific lectures. It was natural that
one who had become interested in physics should
wish to study chemistry. The teacher (who was
not "Uncle Good") had never so much as laid
eyes on a textbook of chemistry; but Edward
was not to be daunted by such trifles. A copy of
Comstock's manual was procured, another pupil
was found willing to join in the study, and this
class of two proceeded to learn what they could
from reading the book, while the teacher asked
them the printed questions, — those questions the
mere existence of which in textbooks is apt to

show what a low view publishers take of the average intelligence of teachers ! It was not a very hopeful way of studying such a subject as chemistry ; but doubtless the time was not wasted, and the foundations for a future knowledge of chemistry were laid. The experience of farm work which accompanied these studies explains the interest which in later years Mr. Youmans felt in agricultural chemistry. He came to realize how crude and primitive are our methods of making the earth yield its produce, and it was his opinion that when men have once learned how to conduct agriculture upon sound scientific principles, farming will become at once the most wholesome and the most attractive form of human industry.

Along with the elementary studies in science there went a great deal of miscellaneous reading, mostly, it would appear, of good solid books. Apparently there was at that time no study of languages, ancient or modern. At the age of seventeen the young man had shown so much promise that it was decided he should study law, and he had already entered upon a more extensive course of preparation in an academy in Saratoga County when the event occurred which changed the whole course of his life. He had been naturally gifted with keen and accurate vision, was a good sports-

man and an excellent shot with a rifle; but at
about the age of thirteen there had come an attack
of ophthalmia, which left the eyes weak and sensitive. Perpetual reading probably increased the
difficulty and hindered complete recovery. At the
age of seventeen violent inflammation set in;
the sight in one eye was completely lost, while in
the other it grew so dim as to be of little avail.
Sometimes he would be just able to find his way
about the streets, at other times the blindness was
almost total; and this state of things lasted for
nearly thirteen years.

This dreadful calamity seemed to make it impossible to continue any systematic course of study,
and the outlook for satisfactory work of any sort
was extremely discouraging. The first necessity
was medical assistance, and in quest of this Mr.
Youmans came in the autumn of 1839 to New
York, where for the most part he spent the remainder of his life. Until 1851 he was under the
care of an oculist. Under such circumstances, if
a man of eager energy and boundless intellectual
craving were to be overwhelmed with despondency,
we could not call it strange. If he were to become dependent upon friends for the means of support, it would be ungracious, if not unjust, to blame
him. But Edward Youmans was not made of the

stuff that acquiesces in defeat. He rose superior
to calamity; he won the means of livelihood, and
in darkness entered upon the path to an enviable
fame. At first he had to resign himself to spending weary weeks over tasks that with sound eye-
sight could have been dispatched in as many days.
He invented some kind of writing machine, which
held his paper firmly, and enabled his pen to follow straight lines at proper distances apart. Long
practice of this sort gave his handwriting a peculiar character which it retained in later years.
When I first saw it in 1863 it seemed almost un-
decipherable; but that was far from being the
case, and after I had grown used to it I found it
but little less legible than the most beautiful chiro-
graphy. The strokes, gnarled and jagged as they
were, had a method in their madness, and every
pithy sentence went straight as an arrow to its
mark.

While conquering these physical obstacles Mr.
Youmans began writing for the press, and gradually entered into relations with leading newspapers which became more and more important
and useful as years went on. He became acquainted with Horace Greeley, William Henry
Channing, and other gentlemen who were interested in social reforms. His sympathies were

strongly enlisted with the little party of abolition-
ists, then held in such scornful disfavour by all
other parties.   He was also interested in the party
of temperance, which, as he and others were afterward to learn, compounded for its essential up-
rightness of purpose by indulging in very gross
intemperance of speech and action.   The disinterestedness which always characterized him was
illustrated by his writing many articles for a temperance paper which could not afford to pay its
contributors, although he was struggling with such
disadvantages in earning his own livelihood and
carrying on his scientific studies.   Those were days
when leading reformers believed that by some cun-
ningly contrived alteration of social arrangements
our human nature, with all its inheritance from
countless ages of brutality, can somehow be made
over all in a moment, just as one would go to work
with masons and carpenters and revamp a house.
There are many good people who still labour under
such a delusion.

Though Mr. Youmans was brought into frequent
contact with reformers of this sort, it does not seem
to me that his mind was ever deeply impressed with
such ways of thinking.   Science is teaching us that
the method of evolution is that mill of God, of
which we have heard, which, while it grinds with

infinite efficacy, yet grinds with wearisome slowness. It was Mr. Darwin's discovery of natural selection which first brought this truth home to us; but Sir Charles Lyell had in 1830 shown how enormous effects are wrought by the cumulative action of slight and unobtrusive causes, and this had much to do with turning men's minds toward some conception of evolution. It was about 1847 that Mr. Youmans was deeply interested in the work of geologists, as well as in the Nebular Theory, to which recent discoveries were adding fresh confirmation. Some time before this he had read that famous book " Vestiges of Creation," and although Professor Agassiz truly declared that it was an unscientific book, crammed with antiquated and exploded fancies, I suspect that Mr. Youmans felt that amid all the chaff there was a very sound and sturdy kernel of truth.

Among the books which Mr. Youmans projected at this time, the first was a compendious history of progress in discovery and invention; but, after he had made extensive preparations, a book was published so similar in scope and treatment that he abandoned the undertaking. Another work was a treatise on arithmetic, on a new and philosophical plan; but, when this was approaching completion, he again found himself anticipated, this

tine by the book of Horace Mann. This was discouraging enough, but a third venture resulted in
a brilliant success. We have observed the eagerness with which, as a schoolboy, Mr. Youmans
entered upon the study of chemistry. His interest
in this science grew with years, and he devoted
himself to it so far as was practicable. For a
blind man to carry on the study of a science which
is preëminently one of observation and experiment
might seem hopeless. It was at any rate absolutely necessary to see with the eyes of others, if
not with his own. Here the assistance rendered
by his sister was invaluable. During most of this
period she served as amanuensis and reader for
him. But, more than this, she kept up for some
time a course of laboratory work, the results of
which were minutely described to her brother and
discussed with him in the evenings. The lectures
of Dr. John William Draper on chemistry were
also thoroughly discussed and pondered.

The conditions under which Mr. Youmans worked
made it necessary for him to consider every point
with the extreme deliberation involved in framing
distinct mental images of things and processes
which he could not watch with the eye. It was
hard discipline, but he doubtless profited from it.
Nature had endowed him with an unusually clear

head, but this enforced method must have made it still clearer. One of the most notable qualities of his mind was the absolute luminousness with which he saw things and the relations among things. It was this quality that made him so successful as an expounder of scientific truths. In the course of his pondering over chemical facts which he was obliged to take at second hand, it occurred to him that most of the pupils in common schools who studied chemistry were practically no better off. It was easy enough for schools to buy textbooks, but difficult for them to provide laboratories and apparatus; and it was much easier withal to find teachers who could ask questions out of a book than those who could use apparatus if provided. It was customary, therefore, to learn chemistry by rote; or, in other words, pupils' heads were crammed with unintelligible statements about things with queer names, — such as manganese or tellurium, — which they had never seen, and would not know if they were to see them. It occurred to Mr. Youmans that if visible processes could not be brought before pupils, at any rate the fundamental conceptions of chemistry might be made clear by means of diagrams. He began devising diagrams in different colours, to illustrate the diversity in the atomic weights of the principal elements,

and the composition of the more familiar com-
pounds. At length, by uniting his diagrams, he
obtained a comprehensive chart exhibiting the outlines of the whole scheme of chemical combination
according to the binary or dualist theory then in
vogue. This chart, when published, was a great
success. It not only facilitated the acquirement
of clear ideas, but it was suggestive of new ideas.
It proved very popular, and kept the field until the
binary theory was overthrown by the modern doctrine of substitution, which does not lend itself so
readily to graphic treatment.

The success of the chemical chart led to the writing of a textbook of chemistry. This laborious
work was completed in 1851, when Mr. Youmans
was thirty years old. Professor Silliman was then
regarded as one of our foremost authorities in chemistry, but it was at once remarked of the new book
that it showed quite as thorough a mastery of the
whole subject of chemical combination as Silliman's. It was a textbook of a kind far less common then than now. There was nothing dry about
it. The subject was presented with beautiful clearness, in a most attractive style. There was a firm
grasp of the philosophical principles underlying
chemical phenomena, and the meaning and functions of the science were set forth in such a way

as to charm the student and make him wish for more. The book had an immediate and signal success; in after years it was twice rewritten by the author, to accommodate it to the rapid advances made by the science, and it is still one of our best textbooks of chemistry. It has had a sale of about one hundred and fifty thousand copies.

The publication of this book at once established its author's reputation as a scientific writer, and in another way it marked an era in his life. The long, distressing period of darkness now came to an end. Sight was so far recovered in one eye that it became possible to go about freely, to read, to recognize friends, to travel, and make much of life. I am told that his face had acquired an expression characteristic of the blind, but that expression was afterward completely lost. When I knew him it would never have occurred to me that his sight was imperfect, except perhaps as regards length of range.

Youmans' career as a scientific lecturer now began. His first lecture was the beginning of a series on the relations of organic life to the atmosphere. It was illustrated with chemical apparatus, and was given in a private room in New York to an audience which filled the room. Probably no lecturer ever faced his first audience without some trepidation, and Youmans had not the mainstay and refuge afforded by a manuscript, for his sight was never good enough to make such an aid available for his lectures. At first the right words were slow in finding their way to those ready lips, and his friends were beginning to grow anxious, when all at once a happy accident broke the spell. He was remarking upon the characteristic inertness of nitrogen, and pointing to a jar of that gas on the table before him, when some fidgety movement of his knocked the jar off the table. He improved the occasion with one of his quaint *bons mots*, and, as there is nothing that greases the wheels of life like a laugh, the lecture went on to a successful close.

This was the beginning of a busy career of seventeen years of lecturing, ending in 1868 ; and I believe it is safe to say that few things were done in all those years of more vital and lasting benefit to the American people than this broadcast sowing of the seeds of scientific thought in the lectures of Edward Youmans. They came just at the time when the world was ripe for the doctrine of evolution, when all the wondrous significance of the trend of scientific discovery since Newton's time was beginning to burst upon men's minds. The work of Lyell in geology, followed at length in

1859 by the Darwinian theory; the doctrine of the correlation of forces and the consequent unity of nature; the extension and reformation of chemical theory; the simultaneous advance made in sociological inquiry, and in the conception of the true aims and proper methods of education, — all this made the period a most fruitful one for the peculiar work of such a teacher as Youmans. The intellectual atmosphere was charged with conceptions of evolution. Youmans had arrived at such conceptions in the course of his study of the separate lines of scientific speculation which were now about to be summed up and organized by Herbert Spencer into that system of philosophy which marks the highest point to which the progressive intelligence of mankind has yet attained. In the field of scientific generalization upon this great scale, Youmans was not an originator; but his broadly sympathetic and luminous mind moved on a plane so near to that of the originators that he seized at once upon the grand scheme of thought as it was developed, made it his own, and brought to its interpretation and diffusion such a happy combination of qualities as one seldom meets with. The ordinary popularizer of great and novel truths is a man who comprehends them but partially, and illustrates them in a lame and frag-

mentary way. But it was the peculiarity of Youmans that while on the one hand he could grasp
the newest scientific thought so surely and firmly
that he seemed to have entered into the innermost
mind of its author, on the other hand he could
speak to the general public in an extremely convincing and stimulating way. This was the secret
of his power, and there can be no question that
his influence in educating the American people
to receive the doctrine of evolution was great and
widespread.

The years when Youmans was travelling and lecturing were the years when the old lyceum system
of popular lectures was still in its vigour. The
kind of life led by the energetic lecturer in those
days was not that of a Sybarite, as may be seen
from a passage in one of his letters: " I lectured
in Sandusky, and had to get up at five o'clock to
reach Elyria; I had had but very little sleep.
To get from Elyria to Pittsburg I must take the
five o'clock morning train, and the hotel darky
said he would *try* to waken me. I knew what
that meant, and so did not get a single wink of
sleep that night. Rode all day to Pittsburg, and
had to lecture in the great Academy of Music over
footlights. . . . The train that left for Zanesville
departed at two in the morning. I had been as-

sured a hundred times (for I asked everybody I met) that I could get a sleeping-car to Zanesville, and when I was all ready to start I was informed that *this* morning there was no sleeping-car. By the time I reached here I was pretty completely used up."

Such a fatiguing life, however, has its compensations. It brings the lecturer into friendly contact with the brightest minds among his fellow countrymen in many and many places, and enlarges his sphere of influence in a way that is not easy to estimate. Clearly, an earnest lecturer, of commanding intelligence and charming manner, with a great subject to teach, must have an opportunity for sowing seeds that will presently ripen in a change of opinion or sentiment, in an altered way of looking at things on the part of whole communities. No lecturer has ever had a better opportunity of this sort than Edward Youmans, and none ever made a better use of his opportunity. His gifts as a talker were of the highest order. The commonest and plainest story, as told by Edward Youmans, had all the breathless interest of the most thrilling romance. Absolutely unconscious of himself, simple, straightforward, and vehement, wrapped up in his subject, the very embodiment of faith and enthusiasm, of heartiness and good

cheer, it was delightful to hear him. And when
we join with all this his unfailing common sense,
his broad and kindly view of men and things, and
the delicious humour that kept flashing out in
quaint, pithy phrases such as no other man would
have thought of, and such as are the despair of
any one trying to remember and quote them, we can
seem to imagine what a power he must have been
with his lectures.

When such a man goes about for seventeen
years, teaching scientific truths for which the world
is ripe, we may be sure that his work is great,
albeit we have no standard whereby we can exactly
measure it. In hundreds of little towns with queer
names did this strong personality appear and make
its way and leave its effects in the shape of new
thoughts, new questions, and enlarged hospitality
of mind, among the inhabitants. The results of
all this are surely visible to-day. In no part of the
English world has Herbert Spencer's philosophy
met with such a general and cordial reception as in
the United States. This may no doubt be largely
explained by a reference to general causes; but as
it is almost always necessary, along with our general causes, to take into the account some personal
influence, so it is in this case. It is safe to say
that among the agencies which during the past

fifty years have so remarkably broadened the mind
of the American people, very few have been more
potent than the gentle and subtle but pervasive
work done by Edward Youmans with his lectures,
and to this has been largely due the hospitable
reception of Herbert Spencer's ideas.

It was in 1856 that Youmans fell in with a
review of "Spencer's Principles of Psychology,"
by Dr. Morell, in "The Medico-Chirurgical Review." This paper impressed him so deeply that
he at once sent to London for a copy of the book,
which had been published in the preceding year.
It will be observed that this was four years before
the Darwinian theory was announced in the first
edition of the "Origin of Species." [1]

After struggling for a while with the weighty
problems of this book, Youmans saw that the theory
expounded in it was a long stride in the direction
of a general theory of evolution. His interest in
this subject received a new and fresh stimulus.
He read "Social Statics," and began to recognize
Spencer's hand in the anonymous articles in the
quarterlies in which he was then announcing and
illustrating various portions or segments of his
newly discovered law of evolution. One evening
in February, 1860, as Youmans was calling at a

[1] See above, p. 49.

friend's house in Brooklyn, the Rev. Samuel Johnson, of Salem, handed him the famous prospectus
of the great series of philosophical works which
Spencer proposed to issue by subscription.   Mr.
Johnson had obtained this from Edward Silsbee,
who was one of the very first Americans to become
interested in Spencer.   The very next day Youmans wrote a letter to Spencer, offering his aid in
procuring American subscriptions and otherwise
facilitating the enterprise by every means in his
power.   With this letter and Spencer's cordial
reply began the lifelong friendship between the
two men.   It was in that same month that I first
became aware of Spencer's existence, through a
single paragraph quoted from him by Lewes, and in
that paragraph there was immense fascination.   I
had been steeping myself in the literature of modern philosophy, starting with Bacon and Descartes,
and was then studying Comte's " Philosophie Positive," which interested me as suggesting that the
special doctrines of the several sciences might be
organized into a general body of doctrine of universal significance.   Comte's work was crude and
often wildly absurd, but there was much in it that
was very suggestive.   In May, 1860, in the Old
Corner Bookstore in Boston, I fell upon a copy of
that same prospectus of Spencer's works, and read

it with exulting delight; for clearly there was to be such an organization of scientific doctrine as the world was waiting for. It appeared that there was some talk of Ticknor & Fields undertaking to conduct the series in case subscriptions enough should be received. Spencer preferred to have his works appear in Boston; but when in the course of 1860 his book on "Education" was offered to Ticknor & Fields, they declined to publish it, — which was, of course, a grave mistake from the business point of view. Youmans, however, was not sorry for this, for it gave him the opportunity to place Spencer's books where he could do most to forward their success.

Some years before, during his blindness, his sister had led him one day into the store of Messrs. D. Appleton & Co. in quest of a book, and Mr. William Appleton had become warmly interested in him. I believe the firm now look back to this chance visit as one of the most auspicious events in their annals. Youmans became by degrees a kind of adviser as regarded matters of publication, and it was largely through his far-sighted advice that the Appletons entered upon the publication of such books as those of Buckle, Darwin, Huxley, Tyndall, Haeckel, and others of like character; always paying a royalty to the authors,

the same as to American authors, in spite of the
absence of an international copyright law. As
publishers of books of this sort the Appletons have
come to be preëminent. It is obvious enough
nowadays that such books are profitable from a
business point of view; but thirty years and more
ago this was by no means obvious. We Americans were terribly provincial. Reprints of English
books and translations from French and German
were sadly behind the times. In the Connecticut
town where I lived, people would begin to wake
up to the existence of some great European book
or system of thought after it had been before the
world anywhere from a dozen to fifty years. In
those days, therefore, it required some boldness to
undertake the reprinting of new scientific books;
and none have recognized more freely than the
Appletons the importance of the part played by
Youmans in this matter. His work as adviser to
a great publishing house and his work as lecturer
reinforced each other, and thus his capacity for
usefulness was much increased.

When Spencer's book on "Education" failed
to find favour in Boston the Appletons took it,
and thus presently secured the management of the
philosophical series. This brought Youmans into
permanent relations with Spencer and his work.

In 1861 Youmans was married, and in the course
of the following year made a journey in Europe
with his wife. It was now that he became per-
sonally acquainted with Spencer, and found him
quite as interesting and admirable as his books.
Friendships were also begun with Huxley and
other foremost men of science. From more than
one of these men I have heard the warmest expressions of personal affection for Youmans, and of
keen appreciation of the aid that they have obtained in innumerable ways from his intelligent
and enthusiastic sympathy. But no one else got
so large a measure of this support as Spencer. As
fast as his books were republished Youmans wrote
reviews of them, and by no means in the usual
perfunctory way; his reviews and notices were
turned out by the score, and scattered about in
the magazines and newspapers where they would
do the most good. Whenever he found another
writer who could be pressed into the service, he
would give him Spencer's books, kindle him with a
spark from his own magnificent enthusiasm, and
set him to writing for the press. The most inde-
fatigable vender of wares was never more ruthlessly
persistent in advertising for lucre's sake than
Edward Youmans in preaching in a spirit of the
purest disinterestedness the gospel of evolution.

As long as he lived, Spencer had upon this side of the Atlantic an *alter ego* ever on the alert with vision like that of a hawk for the slightest chance to promote his interests and those of his system of thought.

Among the allies thus enlisted at that early time were Mr. George Ripley and the Rev. Henry Ward Beecher, both of whom did good service, in their different ways, in awakening public interest in the doctrine of evolution. In those days of the Civil War it was especially hard to keep up the list of subscribers in an abstruse philosophical publication of apparently interminable length. Youmans now and then found it needful to make a journey in the interests of the work, and it was on one of these occasions, in November, 1863, that I made his acquaintance. I had already published, in 1861, an article in one of the quarterly reviews, in which Spencer's work was referred to; and another in 1863, in which the law of evolution was illustrated in connection with certain problems of the science of language. The articles were anonymous, as was then the fashion, and Youmans' curiosity was aroused. There were so few people then who had any conception of what Spencer's work meant that they could have been counted on one's fingers. At that time I knew of only three: the late Professor Gurney, of Harvard; Mr. George Litch Roberts, now an eminent patent lawyer in Boston; and Mr. John Spencer Clark, now of the Prang Educational Company. I have since known that there were at least two or three others about Boston, among them my learned friend the Rev. William Rounseville Alger, besides several in other parts of the country. When we sometimes ventured to observe that Spencer's work was as great as Newton's, and that his theory of evolution was going to remodel human thinking upon all subjects whatever, people used to stare at us and take us for idiots. Any one member of such a small community was easy to find; and I have always dated a new era in my life from the Sunday afternoon when Youmans came to my room in Cambridge. It was the beginning of a friendship such as hardly comes but once to a man. At that first meeting I knew nothing of him except that he was the author of a textbook of chemistry which I had found interesting, in spite of its having been crammed down my throat by an old-fashioned memorizing teacher who, I am convinced, never really knew so much as the difference between oxygen and antimony. At first it was a matter of breathless interest to talk with a man who had seen Herbert Spencer. But one of the immediate results of this interview was the beginning of my own correspondence and intimate friendship with Spencer. And from that time forth it always seemed as if, whenever any of the good or lovely things of life came to my lot, somehow or other Edward Youmans was either the cause of it, or at any rate intimately concerned with it. The sphere of his unselfish goodness was so wide and its quality so potent that one could not come into near relations with him without becoming in all manner of unsuspected ways strengthened and enriched.

In the autumn of 1865 we were dismayed by the announcement that Spencer would no longer be able to go on issuing his works. In London they were published at his own expense and risk, and those books which now yield a handsome profit did not then pay the cost of making them. By the summer of 1865 there was a balance of £1100 against Spencer, and his property was too small to admit of his going on and losing at such a rate. As soon as this was known, John Stuart Mill begged to be allowed to assume the entire pecuniary responsibility of continuing the publication; but this, Mr. Spencer, while deeply affected by such noble sympathy, would not hear of. He consented, however, with great reluctance, to the attempt of Huxley and Lubbock, and other friends, to increase artificially the list of subscribers by inducing people to take the work just in order to help support it. But after several months the sudden death of Spencer's father added something to his means of support, and he thereupon withdrew his consent to this arrangement, and determined to go on publishing as before, and bearing the loss.

But as soon as the first evil tidings reached America Youmans made up his mind that $5500 must be forthwith raised by subscription, in order to make good the loss already incurred. It is delightful to remember the vigour with which he took hold of this work. The sum of $7000 was raised and invested in American securities in Spencer's name. If he did not see fit to accept these securities, they would go without an owner. The best of Waltham watches was procured for Spencer by his American friends; a letter, worded with rare delicacy and tact, was written by the late Robert Minturn; and Youmans sailed for England to convey the letter and the watch to Spencer. It was a charming scene on a summer day in an English garden when the great philosopher was apprised of what had been done. It was so skilfully managed that he could not refuse the tribute without seeming churlish. He therefore accepted it, and applied it to extending his researches in descriptive sociology.

Of the many visits which Youmans made to England, now and then extending them to the Continent, one of the most important was in 1871, for the purpose of establishing the International Scientific Series. This was a favourite scheme of Youmans. He realized that popular scientific books, adapted to the general reader, are apt to be written by third-rate men who do not well understand their subject ; they are apt to be dry or superficial, or both. No one can write so good a popular book as the master of a subject, if he only has a fair gift of expressing himself and keeps in mind the public for which he is writing. The master knows what to tell and what to omit, and can thus tell much in a short compass and still make it interesting ; moreover, he avoids the inaccuracies which are sure to occur in second-hand work. Masters of subjects are apt, however, to be too much occupied with original research to write popular books. It was Youmans' plan to induce the leading men of science in Europe and America to contribute small volumes on their special subjects to a series to be published simultaneously in several countries and languages.

Furthermore, by special contract with publishing houses of high reputation, the author was to receive the ordinary royalty on every copy of his book sold in every one of the countries in question ; thus anticipating international copyright upon a very wide scale, and giving the author a much more adequate compensation for his labour. To put this scheme into operation was a task of great difficulty, so many conflicting interests had to be considered. Youmans' brilliant success is attested by that noble series of more than fifty volumes, on all sorts of scientific subjects, written by men of real eminence, and published in England, France, Italy, Germany, and Russia, as well as in the United States.

A word is all that can be spared for other parts of our friend's work, which deserve many words, and those carefully considered. His book on " Household Science " is not the usual collection of scrappy comment, recipe, and apothegm, but a valuable scientific treatise on heat, light, air, and food in their relations to every-day life. In his " Correlation of Physical Forces " he brings together the epoch-making essays of the men who have successively established that doctrine, introducing them with an essay of his own, in which its history and its philosophical implications are set forth in a masterly manner. In his book on the "Culture demanded by Modern Life" we have a similar collection of essays with a similar excellent original discussion, showing the need for wider and later training in science, and protesting against the excess of time and energy that is spent in classical education where it is merely the following of an old tradition.

As a crown to all this useful work, Youmans established, in 1872, "The Popular Science Monthly," which has unquestionably been of high educational value to the general public. It was not the aim of this magazine to give an account of every theory expounded, every fact observed, every discovery made, from year to year, whether significant or insignificant. The mind of the people is not educated by dumping a great unshapely mass of facts into it. It needs to be stimulated rather than crammed. Education in science should lead one to think for one's self. The scientific magazine, therefore, should present articles from all quarters that deal with the essential conceptions of science or discuss problems of real theoretical or practical interest, no matter whether every particular asteroid or the last new species of barnacle receives full attention or not. "The Popular Science Monthly" has now been with us eighteen years ; its character has always been of the highest, and it must have exerted an excellent influence not only as a diffuser of valuable knowledge, but in training its readers to scientific habits of thought in so far as mere reading can contribute to such a result.

In concluding our survey of this useful and noble life, what impresses us most, I think, is the broad democratic spirit and the absolute unselfishness which it reveals at every moment and in every act. To Edward Youmans the imperative need for educating the great mass of the people so as to use their mental powers to the best advantage came home as a living, ever present fact. He saw all that it meant and means in the raising of mankind to a higher level of thought and action than that upon which they now live. To this end he consecrated himself with unalloyed devotion ; and we who mourn his loss look back upon his noble career with a sense of victory, knowing how the good that such a man does lives after him and can never die.

March, 1890.

IV

THE PART PLAYED BY INFANCY IN THE EVOLUTION OF MAN [1]

THE remarks which my friend Mr. Clark has made with reference to the reconciling of science and religion seem to carry me back to the days when I first became acquainted with the fact that there were such things afloat in the world as speculations about the origin of man from lower forms of life; and I can recall step by step various stages in which that old question has come to have a different look from what it had thirty years ago. One of the commonest objections we used to hear, from the mouths of persons who could not very well give voice to any other objection, was that anybody, whether he knows much or little about evolution, must have the feeling that there is something degrading about being allied with lower forms of life. That was, I suppose, owing to the survival of the

---

[1] Short-hand report of my speech at a dinner given for me by Mr. John Spencer Clark, at the Aldine Club, New York, May 13, 1895.

old feeling that a dignified product of creation ought
to have been produced in some exceptional way.
That which was done in the ordinary way, that
which was done through ordinary processes of cau-
sation, seemed to be cheapened and to lose its value.
It was a remnant of the old state of feeling which
took pleasure in miracles, which seemed to think
that the object of thought was more dignified if
you could connect it with something supernatural;
that state of culture in which there was an altogether inadequate appreciation of the amount of
grandeur that there might be in the slow creative
work that goes on noiselessly by little minute in-
crements, even as the dropping of the water that
wears away the stone. The general progress of
familiarity with the conception of evolution has
done a great deal to change that state of mind.
Even persons who have not much acquaintance
with science have at length caught something of its
lesson, — that the infinitely cumulative action of
small causes like those which we know is capable of
producing results of the grandest and most thrilling importance, and that the disposition to recur to
the cataclysmic and miraculous is only a tendency
of the childish mind which we are outgrowing
with wider experience.

The whole doctrine of evolution, and in fact the

whole advance of modern science from the days of
Copernicus down to the present day, have consisted in the substitution of processes which are
familiar and the application of those processes,
showing how they produce great results.

When Darwin's " Origin of Species " was first
published, when it gave us that wonderful explanation of the origin of forms of life from allied forms
through the operation of natural selection, it must
have been like a mental illumination to every person who comprehended it.   But after all it left a
great many questions unexplained, as was natural.
It accounted for the phenomena of organic development in general with wonderful success, but it must
have left a great many minds with the feeling: If
man has been produced in this way, if the mere
operation of natural selection has produced the
human race, wherein is the human race anyway
essentially different from lower races?   Is not
man really dethroned, taken down from that exceptional position in which we have been accustomed
to place him, and might it not be possible, in the
course of the future, for other beings to come upon
the earth as far superior to man as man is superior
to the fossilized dragons of Jurassic antiquity ?

Such questions used to be asked, and when they
were asked, although one might have a very strong

feeling that it was not so, at the same time one could not exactly say why. One could not then find any scientific argument for objections to that point of view. But with the further development of the question the whole subject began gradually to wear a different appearance; and I am going to give you a little bit of autobiography, because I think it may be of some interest in this connection. I am going to mention two or three of the successive stages which the whole question took in my own mind as one thing came up after another, and how from time to time it began to dawn upon me that I had up to that point been looking at the problem from not exactly the right point of view.

When Darwin's " Descent of Man " was published in 1871, it was of course a book characterized by all his immense learning, his wonderful fairness of spirit and fertility of suggestion. Still, one could not but feel that it did not solve the question of the origin of man. There was one great contrast between that book and his " Origin of Species." In the earlier treatise he undertook to point out a *vera causa* of the origin of species, and he did it. In his " Descent of Man " he brought together a great many minor generalizations which facilitated the understanding of man's origin. But he did not come at all near to solving the central problem, nor did he anywhere show clearly why natural selection might not have gone on forever producing one set of beings after another distinguishable chiefly by physical differences. But Darwin's co-discoverer, Alfred Russel Wallace, at an early stage in his researches, struck out a most brilliant and pregnant suggestion. In that one respect Wallace went further than ever Darwin did. It was a point of which, indeed, Darwin admitted the importance. It was a point of which nobody could fail to understand the importance, that in the course of the evolution of a very highly organized animal, if there came a point at which it was of more advantage to that animal to have variations in his intelligence seized upon and improved by natural selection than to have physical changes seized upon, then natural selection would begin working almost exclusively upon that creature's intelligence, and he would develop in intelligence to a great extent, while his physical organism would change but slightly. Now, that of course applied to the case of man, who is changed physically but very slightly from the apes, while he has traversed intellectually such a stupendous chasm.

As soon as this statement was made by Wallace, it seemed to me to open up an entirely new world of speculation. There was this enormous antiquity

of man, during the greater part of which he did
not know enough to make history. We see man
existing here on the earth, no one can say how
long, but surely many hundreds of thousands of
years, yet only during just the last little fringe of
four or five thousand years has he arrived at the
point where he makes history. Before that, something was going on, a great many things were going
on, while his ancestors were slowly growing up to
that point of intelligence where it began to make
itself felt in the recording of events. This agrees
with Wallace's suggestion of a long period of psychical change, accompanied by slight physical change.

Well, in the spring of 1871, when Darwin's
" Descent of Man " came out, just about the same
time I happened to be reading Wallace's account
of his experiences in the Malay Archipelago,
and how at one time he caught a female orang-
outang with a new-born baby, and the mother died,
and Wallace brought up the baby orang-outang
by hand ; and this baby orang-outang had a kind
of infancy which was a great deal longer than that
of a cow or a sheep, but it was nothing compared
to human infancy in length. This little orang-
outang could not get up and march around, as
mammals of less intelligence do, when he was first
born, or within three or four days ; but after three

or four weeks or so he would get up, and begin
taking hold of something and pushing it around,
just as children push a chair; and he went through
a period of staring at his hands, as human babies
do, and altogether was a good deal slower in getting to the point where he could take care of himself.   And while I was reading of that I thought,
Dear me! if there is any one thing in which the
human race is signally distinguished from other
mammals, it is in the enormous duration of their
infancy; but it is a point that I do not recollect
ever seeing any naturalist so much as allude to.

It happened at just that time that I was making researches in psychology about the organization
of experiences, the way in which conscious intelligent action can pass down into quasi-automatic
action, the generation of instincts, and various
allied questions; and I thought, Can it be that the
increase of intelligence in an animal, if carried
beyond a certain point, must necessarily result in
prolongation of the period of infancy, — must
necessarily result in the birth of the mammal at a
less developed stage, leaving something to be done,
leaving a good deal to be done, after birth?   And
then the argument seemed to come along very
naturally, that for every action of life, every adjustment which a creature makes in life, whether a

muscular adjustment or an intelligent adjustment, there has got to be some registration effected in the nervous system, some line of transit worn for nervous force to follow; there has got to be a connection between certain nerve-centres before the thing can be done, whether it is the acts of the viscera or the acts of the limbs, or anything of that sort; and of course it is obvious that if the creature has not many things to register in his nervous system, if he has a life which is very simple, consisting of few actions that are performed with great frequency, that animal becomes almost automatic in his whole life; and all the nervous connections that need to be made to enable him to carry on life get made during the fœtal period or during the egg period, and when he comes to be born, he comes all ready to go to work. As one result of this, he does not learn from individual experience, but one generation is like the preceding generations, with here and there some slight modifications. But when you get the creature that has arrived at the point where his experience has become varied, he has got to do a good many things, and there is more or less individuality about them; and many of them are not performed with the same minuteness and regularity, so that there does not begin to be that automatism within the period during which he is

being developed and his form is taking on its outlines.   During prenatal life there is not time
enough for all these nervous registrations, and so
by degrees it comes about that he is born with his
nervous system perfectly capable only of making
him breathe and digest food, — of making him do
the things absolutely requisite for supporting life ;
instead of being born with a certain number of
definite developed capacities, he has a number of
potentialities which have got to be roused according to his own individual experience.   Pursuing
that line of thought, it began after a while to seem
clear to me that the infancy of the animal in a very
undeveloped condition, with the larger part of his
faculties in potentiality rather than in actuality,
was a direct result of the increase of intelligence,
and I began to see that now we have two steps :
first, natural selection goes on increasing the intelligence ; and secondly, when the intelligence goes
far enough, it makes a longer infancy, a creature
is born less developed, and therefore there comes
this plastic period during which he is more teachable.   The capacity for progress begins to come
in, and you begin to get at one of the great
points in which man is distinguished from the
lower animals, for one of those points is undoubtedly his progressiveness ; and I think that any one

will say, with very little hesitation, that if it were not for our period of infancy we should not be progressive. If we came into the world with our capacities all cut and dried, one generation would be very much like another.

Then, looking round to see what are the other points which are most important in which man differs from the lower animals, there comes that matter of the family. The family has adumbrations and foreshadowings among the lower animals, but in general it may be said that while mammals lower than man are gregarious, in man have become established those peculiar relationships which constitute what we know as the family; and it is easy to see how the existence of helpless infants would bring about just that state of things. The necessity of caring for the infants would prolong the period of maternal affection, and would tend to keep the father and mother and children together, but it would tend especially to keep the mother and children together. This business of the marital relations was not really a thing that became adjusted in the primitive ages of man, but it has become adjusted in the course of civilization. Real monogamy, real faithfulness of the male parent, belongs to a comparatively advanced stage; but in the early stages the knitting together of permanent relations between mother and infant, and the approximation toward steady relations on the part of the male parent, came to bring about the family, and gradually to knit those organizations which we know as clans.

Here we come to another stage, another step forward. The instant society becomes organized in clans, natural selection cannot let these clans be broken up and die out, — the clan becomes the chief object or care of natural selection, because if you destroy it you retrograde again, you lose all you have gained ; consequently, those clans in which the primeval selfish instincts were so modified that the individual conduct would be subordinated to some extent to the needs of the clan, — those are the ones which would prevail in the struggle for life. In this way you gradually get an external standard to which man has to conform his conduct, and you get the germs of altruism and morality ; and in the prolonged affectionate relation between the mother and the infant you get the opportunity for that development of altruistic feeling which, once started in those relations, comes into play in the more general relations, and makes more feasible and more workable the bonds which keep society together, and enable it to unite on wider and wider terms.

So it seems that from a very small beginning we

are reaching a very considerable result. I had got these facts pretty clearly worked out, and carried them around with me some years, before a fresh conclusion came over me one day with a feeling of surprise. In the old days before the Copernican astronomy was promulgated, man regarded himself as the centre of the universe. He used to entertain theological systems which conformed to his limited knowledge of nature. The universe seemed to be made for his uses, the earth seemed to have been fitted up for his dwelling place, he occupied the centre of creation, the sun was made to give him light, etc. When Copernicus overthrew that view, the effect upon theology was certainly tremendous. I do not believe that justice has ever been done to the shock that it gave to man when he was made to realize that he occupied a kind of miserable little clod of dirt in the universe, and that there were so many other worlds greater than this. It was one of the first great shocks involved in the change from ancient to modern scientific views, and I do not doubt it was responsible for a great deal of the pessimistic philosophizing that came in the seventeenth and eighteenth centuries.

Now, it flashed upon me a dozen years or so ago — after thinking about this manner in which man

originated — that man occupies certainly just as exceptional a position as before, if he is the terminal
in a long series of evolutionary events. If at the
end of the long history of evolution comes man, if
this whole secular process has been going on to
produce this supreme object, it does not much matter what kind of a cosmical body he lives on. He
is put back into the old position of theological importance, and in a much more intelligent way than
in the old days when he was supposed to occupy
the centre of the universe. We are enabled to say
that while there is no doubt of the evolutionary
process going on throughout countless ages which
we know nothing about, yet in the one case where
it is brought home to us we spell out an intelligible
story, and we do find things working along up to
man as a terminal fact in the whole process. This
is indeed a consistent conclusion from Wallace's
suggestion that natural selection, in working toward the genesis of man, began to follow a new
path and make psychical changes instead of physical changes. Obviously, here you are started upon
a new chapter in the history of the universe. It is
no longer going to be necessary to shape new limbs,
and to thicken the skin and make new growths of
hair, when man has learned how to build a fire,
when he can take some other animal's hide and

make it into clothes. You have got to a new state of things.

After I had put together all these additional circumstances with regard to the origination of human society and the development of altruism, I began to see a little further into the matter. It then began to appear that not only is man the terminal factor in a long process of evolution, but in the origination of man there began the development of the higher psychical attributes, and those attributes are coming to play a greater and greater part in the development of the human race. Just take this mere matter of " altruism," as we call it. It is not a pretty word, but must serve for want of a better. In the development of altruism from the low point, where there was scarcely enough to hold the clan together, up to the point reached at the present day, there has been a notable progress, but there is still room for an enormous amount of improvement. The progress has been all in the direction of bringing out what we call the higher spiritual attributes. The feeling was now more strongly impressed upon me than ever, that all these things tended to set the whole doctrine of evolution into harmony with religion; that if the past through which man had originated was such as has been described, then religion was a fit and worthy occupa-

tion for man, and some of the assumptions which
underlie every system of religion must be true.   For
example, with regard to the assumption that what
we see of the present life is not the whole thing ;
that there is a spiritual side of the question beside
the material side ; that, in short, there is for man a
life eternal.   When I wrote the " Destiny of Man,"
all that I ventured to say was, that it did not seem
quite compatible with ordinary common sense to
suppose that so much pains would have been taken
to produce a merely ephemeral result.   But since
then another argument has occurred to me : that
just at the time when the human race was beginning to come upon the scene, when the germs of
morality were coming in with the family, when society was taking its first start, there came into the
human mind — how one can hardly say, but there
did come — the beginnings of a groping after something that lies outside and beyond the world of sense.
That groping after a spiritual world has been
going on here for much more than a hundred thousand years, and it has played an enormous part in
the history of mankind, in the whole development
of human society.   Nobody can imagine what mankind would have been without it up to the present
time.   Either all religion has been a reaching out
for a phantom that does not exist, or a reaching

out after something that does exist, but of which man, with his limited intelligence, has only been able to gain a crude idea. And the latter seems a far more probable conclusion, because, if it is not so, it constitutes a unique exception to all the operations of evolution we know about. As a general thing in the whole history of evolution, when you see any internal adjustment reaching out toward something, it is in order to adapt itself to something that really exists; and if the religious cravings of man constitute an exception, they are the one thing in the whole process of evolution that is exceptional and different from all the rest. And this is surely an argument of stupendous and resistless weight.

I take this autobiographical way of referring to these things, in the order in which they came before my mind, for the sake of illustration. The net result of the whole is to put evolution in harmony with religious thought, — not necessarily in harmony with particular religious dogmas or theories, but in harmony with the great religious drift, so that the antagonism which used to appear to exist between religion and science is likely to disappear. So I think it will before a great while. If you take the case of some evolutionist like Professor Haeckel, who is perfectly sure that materialism

accounts for everything (he has got it all cut and
dried and settled; he knows all about it, so that
there is really no need of discussing the subject!) ;
if you ask the question whether it was his scientific
study of evolution that really led him to such a dogmatic conclusion, or whether it was that he started
from some purely arbitrary assumption, like the
French materialists of the eighteenth century, I
have no doubt the latter would be the true explanation.   There are a good many people who start
on their theories of evolution with these ultimate
questions all settled to begin with.   It was the
most natural thing in the world that after the first
assaults of science upon old beliefs, after a certain number of Bible stories and a certain number
of church doctrines had been discredited, there
should be a school of men who in sheer weariness
should settle down to scientific researches, and say,
" We content ourselves with what we can prove
by the methods of physical science, and we will
throw everything else overboard."   That was very
much the state of mind of the famous French
atheists of the last century.   But only think how
chaotic nature was to their minds compared to
what she is to our minds to-day.   Just think how
we have in the present century arrived where we
can see the bearings of one set of facts in nature

as collated with another set of facts, and contrast it with the view which even the greatest of those scientific French materialists could take. Consider how fragmentary and how lacking in arrangement was the universe they saw compared with the universe we can see to-day, and it is not strange that to them it could be an atheistic world. That hostility between science and religion continued as long as religion was linked hand in hand with the ancient doctrine of special creation. But now that the religious world has unmoored itself, now that it is beginning to see the truth and beauty of natural science and to look with friendship upon conceptions of evolution, I suspect that this temporary antagonism, which we have fallen into a careless way of regarding as an everlasting antagonism, will come to an end perhaps quicker than we realize.

There is one point that is of great interest in this connection, although I can only hint at it. Among the things that happened in that dim past when man was coming into existence was the increase of his powers of manipulation; and that was a factor of immense importance. Anaxagoras, it is said, wrote a treatise in which he maintained that the human race would never have become human if it had not been for the hand. I do not know that there was so very much exaggeration

about that.   It was certainly of great significance
that the particular race of mammals whose intelligence increased far enough to make it worth while
for natural selection to work upon intelligence alone
was the race which had developed hands and could
manipulate things.   It was a wonderful era in the
history of creation when that creature could take
a club and use it for a hammer, or could pry up a
stone with a stake, thus adding one more lever to
the levers that made up his arm.   From that day
to this, the career of man has been that of a person
who has operated upon his environment in a different way from any animal before him.   An era of
similar importance came probably somewhat later,
when man learned how to build a fire and cook his
food; thus initiating that course of culinary development of which we have seen the climax in our
dainty dinner this evening.   Here was another
means of acting upon the environment.   Here was
the beginning of the working of endless physical
and chemical changes through the application of
heat, just as the first use of the club or the crow-
bar was the beginning of an enormous development
in the mechanical arts.

Now, at the same time, to go back once more
into that dim past, when ethics and religion,
manual art and scientific thought, found expres-

sion in the crudest form of myths, the æsthetic
sense was germinating likewise. Away back in
the glacial period you find pictures drawn and
scratched upon the reindeer's antler, portraitures
of mammoths and primitive pictures of the chase;
you see the trinkets, the personal decorations, proving beyond question that the æsthetic sense was
there. There has been an immense æsthetic development since then. And I believe that in the
future it is going to mean far more to us than
we have yet begun to realize. I refer to the kind
of training that comes to mankind through direct
operation upon his environment, the incarnation
of his thought, the putting of his ideas into new
material relations. This is going to exert powerful effects of a civilizing kind. There is something
strongly educational and disciplinary in the mere
dealing with matter, whether it be in the manual
training school, whether it be in carpentry, in over-
coming the inherent and total depravity of inani-
mate things, shaping them to your will, and also
in learning to subject yourself to their will (for
sometimes you must do that in order to achieve
your conquests; in other words, you must humour
their habits and proclivities). In all this there is
a priceless discipline, moral as well as mental, let
alone the fact that, in whatever kind of artistic

work a man does, he is doing that which in the very working has in it an element of something outside of egoism ; even if he is doing it for motives not very altruistic, he is working toward a result the end of which is the gratification or the benefit of other persons than himself ; he is working toward some result which in a measure depends upon their approval, and to that extent tends to bring him into closer relations to his fellow man.

In the future, to an even greater extent than in the recent past, crude labour will be replaced by mechanical contrivances. The kind of labour which can command its price is the kind which has trained intelligence behind it. One of the great needs of our time is the multiplication of skilled and special labour. The demand for the products of intelligence is far greater than that for mere crude products of labour, and it will be more and more so. For there comes a time when the latter products have satisfied the limit to which a man can consume food and drink and shelter, — those things which merely keep the animal alive. But to those things which minister to the requirements of the spiritual side of a man there is almost no limit. The demand one can conceive is wellnigh infinite. One of the philosophical things that have been said, in discriminating man from the lower animals, is that he is the one creature who is never satisfied. It is well for him that he is so, that there is always something more for which he craves. To my mind, this fact most strongly hints that man is infinitely more than a mere animate machine.

May, 1895.

V

THE ORIGINS OF LIBERAL THOUGHT IN AMERICA [1]

In approaching the subject of the origins of liberal thought in America, one cannot help remembering that the discovery of the new continent was itself such a stimulus to free thinking as the world had never before witnessed. From time immemorial, the trade between Europe and the remote parts of Asia had followed certain customary routes. From ancient days, long before Olympiads were heard of, when Assyrian kings with curly beards commemorated their victories in arrow-headed inscriptions, men had used those same routes. Up the Red Sea, in the early prime of hundred-gated Thebes, came ships from the Indian Ocean, with gems and spices to exchange for Egyptian fine linens and amulets of amber from the Baltic; and five thousand years later Venetian argosies at Alexandria were laden with just such

[1] An address delivered at the National Conference of Unitarian Churches, at Washington, D. C., October 23, 1895.

gems and spices to distribute to the merchants of
Augsburg, the royal household at Paris, the lords
and ladies of Haddon Hall.   Empires rose and
fell, creeds and pantheons came and went, stately
temples reared their heads for centuries and slowly
crumbled in ruins, and still amid all the secular
change the world's great stream of trade flowed
through the same unshifting channels, and there
was nothing to show that this state of things, to
which men's ideas and habits had always been adjusted, was not to endure forever.   So it was in
that recent time when Henry V. of England was
smiting the French chivalry at Agincourt, and his
cousin Prince Henry of Portugal was beginning
the search for an ocean route to the Indies.   Never
did the human mind get such a wrench out of its
ancient grooves, never were such vistas of new possibilities laid open, never was beheld such glorious
hardihood, such startling romance, as in the time
when Columbus sailed westward to find the East,
and Cortes met warriors of the Stone Age face to
face.   The men of Europe suddenly found themselves placed in new and unsuspected relations to
the planet on which they lived ; worlds of barbarism and savagery, unheard of and unspeakably
bizarre, were brought to their notice ; strange con-
stellations arose in the firmament ; strange beasts

and birds were encountered amid outlandish trees
and shrubs in new climates beyond unknown seas.
The old familiarity with nature's aspects received
an abrupt shock. On every side loomed up new
questions to be answered, new practical problems
to be solved. All man's inventive faculty, all his
patient inquisitiveness, all the courage he could
summon, were forthwith called into play. The
dreams of boundless riches, the eager thirst for new
knowledge, the superhuman bravery, which characterized the epoch of maritime discovery, are symptoms that reveal to us the highly wrought condition of the European mind at the time. A study
of contemporary chronicles and letters cannot fail
to bring home to us the singular intensity with
which the thrill of venturesome romance was felt
in every fibre of man's being.

The impulse thus given to free thinking must
have been extremely powerful. It is customary
to attribute the brilliant efflorescence of the human mind in the sixteenth century to the revival
of Greek learning. Without seeking to diminish
the respect due to that mighty cause, it may be
contended that the influence of maritime discovery
was equally important. While the Greek renaissance brought men into wholesome and stimulating intercourse with the highest achievements of

literature, art, and philosophy, the discovery of the
New World impressed upon them, as nothing had
ever done before, the feasibleness of doing things
in novel ways. With the wholesale displacement
of commercial relations, the European mind burst
the bounds of the snug little world to which its
habits and theories, its politics civil and ecclesiastical, its science and its theology, had been adapted.
The sudden and unprecedented widening of the
environment soon set up a general fermentation of
ideas. There was nothing accidental in Martin
Luther's coming in the next generation after Columbus. Nor was it strange that in the following
age the English mind, wrought to its highest tension under the combined influences of Renaissance,
Reformation, and maritime adventure, should have
put forth a literature the boldest and grandest
that had ever appeared ; that the era of Raleigh
and Frobisher and the early Puritans should have
seen even the highest mark of Greek achievement
surpassed by Shakespeare. The gigantic revolution set on foot by Copernicus was already in full
progress, the era of Descartes was just arriving,
and the next century was to see modern scientific
method receive its supreme illustration at the
hands of Newton, while the principles of freedom
in thought and speech were to find invincible
champions in Milton and Locke.

Such was the age in which the work of English colonization in America was beginning. In looking for the origins of liberal thought in America, it is chiefly with English-speaking America that we are concerned. The Spanish mind, indeed, felt strongly the stimulus of the maritime discoveries and the contact with strange races of men, until an age of chivalrous enterprise bloomed forth in the literature of Calderon and Lope de Vega and Cervantes; but the new spirit was not strong enough to prevail over an ecclesiastical organization that had been growing in power since the Visigothic times. The higher intellectual life of Spain perished in the fires of the Inquisition, and art and song failed to lead the way to science and free thought; no Spanish Locke or Newton followed in the train of a Lope and a Murillo, but so lately as the year 1771 the University of Salamanca prohibited the teaching of the law of gravitation as discordant with revealed religion.[1] With such a state of things in the mother country, liberal thought in the Spanish colonies was a plant of very slow growth. As for France at the end of the sixteenth century, there was a sturdy intellectual life there which no efforts of tyranny could more than partially repress; but circumstances

[1] Sempere, *Monarchie Espagnole*, ii. 152.

threw the work of colonization into the hands of the Jesuits, and accordingly the history of New France, while eminent for devoted bravery and heroic endurance, shows scarcely a trace of liberal thinking either in politics or in matters pertaining to religion. Not with the French and Spanish portions of America, therefore, but with the colonies that developed into the United States, is our inquiry concerned.

The first and most obvious consideration which strikes us is that while the two centuries following the discovery of America witnessed an unprecedented awakening of the European mind, yet it was only with those nations that had retained self-government that this intellectual awakening was to come to prompt and full fruition. From the British islands and the Netherlands came the kind of public policy that allowed free thinking to take deep root and send up a thrifty tree of liberty. The planting of such seed in the spacious virgin soil of the New World was doubtless the greatest of all the manifold unforeseen results for which Columbus opened the way. It made political freedom the strongest power on earth, and thus favoured the attainment of that equable flexibility of mind which allows the thought to play freely about the facts which are laid before it. Not in

a moment was such a grand result achieved ; its
complete realization has not yet come, and none of
us may live to see it, yet toward that goal the
whole impetus of men's civilizing work is tending,
and there is no power that can prevent the consummation.   Year by year, no matter how grave the
questions with which we have to deal, we are becoming more and more able to let our minds play
freely with them, to turn them hither and thither
till all sides be seen and all aspects duly considered.

Not all in a moment, I say, has such a desirable
result been achieved.   So far is it, moreover, from
having been brought about by conscious human
effort that mankind have in general struggled
desperately against it.   Compared with the mass
of men, it is only a few minds that have learned
to regard absolute freedom of thought as something to be desired.   Though the colonization of
America came at a time when men's minds were
stirred by novel ideas as never before, though the
men of that generation were moving irrepressibly
toward liberality of thought, yet there were very
few who had any liking for liberal thought, or any
good word to bestow upon it.   There were few
who doubted that absolute truth was attainable
concerning the most abstruse questions of philo-

sophy and religion, and an exactly true belief on
minute points of theology was deemed necessary
for one's personal salvation. Changes in opinion
simply wrought a transfer of allegiance from one
orthodoxy to another, and the new orthodoxy felt
bound as much as the old one to persecute all who
refused such allegiance. From this point of view
the history of the progress of liberal thought becomes curiously interesting, for it shows how one
of the most momentous revolutions in human life
has steadily gone on in spite of the inveterate
antagonism of the very men concerned in bringing
it about! To a considerable extent, the history
has been the same over a large part of the globe.
The causes which have been at work in America
have also been at work in Europe, and even beyond; and the liberal thought with which we are
familiar is characteristic not so much of America
as of the latter part of the nineteenth century.
But along with the general causes there have been
local causes which have especially concerned the
New World, and a clear account of the matter
requires us to indicate both the one and the other.

From the revolt of Henry VIII. against the
Papacy down to the Revolution of 1688, there was
in England a progressive movement toward liberal
thought. It was at first a crude unconscious move-

ment in the direction of toleration, which is a necessary condition for the development of free thinking. When we have arrived at a truly cordial
toleration of opinions, allowing to all free play to
stand or fall, just as hypotheses in science are suffered to stand or fall, then is men's thought for the
first time really untrammelled. Whatever, therefore, tended toward toleration of diverse forms of
creed or worship was a step in the path that led
to free thinking; and whatever tended to demo-
cratize the church and relieve it from state control
was a step toward toleration. The revolt of Henry
VIII. at first but realized what the *præmunire*
statutes of Edward I. and Edward III. had threatened. But by breaking up the religious orders,
expelling abbots from Parliament, and making the
headship of the church a subject of fierce dispute,
it contributed immensely to weaken and relax the
bonds of conservatism, and it afforded a rare opportunity for the thoughts of laymen and small
preachers to assert themselves. Thus the Lollard-
ism which had been partially suppressed for more
than a century now reared its head again defi-
antly, and, after learning lessons in democracy
from Calvin, came forth as Puritanism, clad in full
panoply for one of the world's most fateful contests.

In the course of Elizabeth's reign we find this Puritanism taking three different shapes. There were the moderate reformers, whose wish was simply to trim and prune the tree of Episcopacy; and secondly, those who were afterward known as "root and branch" men, whose name is descriptive enough. Instead of pruning they would uproot the tree and cast it away. To these Presbyterians the royal supremacy was no more than the papal a part of the living growth of Christ's church; it was but stubble fit for burning. Kings looked with horror upon such views, which threatened political danger no less than ecclesiastical. "A Scottish presbytery," cried James I., "agreeth as well with a monarchy as God and the Devil. Then Jack and Tom and Will and Dick shall meet, and at their pleasures censure me and my council and all our proceedings." The case could not have been more pithily stated, yet even Presbyterianism stopped short of full-fledged democracy. For Jack and his friends, by means of synods and general assemblies, could create a governing body with power of enforcing conformity upon unwilling congregations. In protest against this somewhat oligarchical method, Puritanism assumed its third form, that of Independency. The beginnings of Independency are to be sought among the Brown-

ists of Elizabeth's reign, though their day of glory
first came with the Civil War. In the theory of
the Independents, as fully developed, any group
of persons wishing to worship God in common
might come together and organize themselves into
a Congregational church, existing by as good a
warrant as any other church, and entirely free
from the control of any bishop, or synod, or council. No outside power could prescribe its creed or
interfere with its ceremonial. Each church became, therefore, a little self-governing republic,
as completely autonomous as an ancient Greek
city, and the union of such churches was based
solely upon the spirit of spontaneous Christian
fellowship. Such was the theory of Independency.

In these successive stages of Protestantism we
may see the preliminary steps toward general toleration and toward liberal thought. In each stage
the strength of the coercive power that could be
exercised over men's opinions and expressions of
opinion was sensibly diminished. From the coercive power of the universal Church, which had
once been able to direct a crusade against the
Albigenses, it was a long step downward to the
coercive power of Queen Elizabeth, whose will to
suppress Puritanism was perpetually held in check

by motives of public policy. It was a yet further
step downward from the coercive power of a sovereign to that of a synod, and thence again to that
of a congregation. So striking is the progress
that one who knew nothing of history might easily
mistake the theory of Independency as providing
practically for something like complete toleration.
History tells us that this was far from being the
case. Heresy, or dissent from the commonly accepted orthodoxy, has been no more tolerated in
Independent churches than elsewhere ; and even
in the absence of serious differences in dogma, persecution has been visited upon divergences from
the customary ritual, as for example in the treatment long accorded to Baptists. In their militant
days, neither Presbyterianism nor Independency
ever professed to be tolerant. The gravest re-
proach they could imagine was to be charged with
encouraging free thinking. The eminent Scottish
divine Rutherford gave expression to the prevail-
ing sentiment when he declared, " We regard toleration of all religions as not far removed from blasphemy." Nevertheless, the movement which gave
rise to Presbyterianism and to Independency was
sure to advance to the announcement of the principle of universal toleration. That movement was
itself the expression of a vast amount of free think-

ing, and it was not to stop short of recognizing
the claims of free thought. The century that
witnessed the beginnings of an English-speaking
America saw also the genuine principles of toleration laid down by Roger Williams and William
Penn, and demonstrated with resistless wealth of
learning and logic by Milton and Locke.

In an account of the origins of liberal thought
in America this English development is all-impor-
tant, but it does not cover the whole field. America's inheritance from Europe comes chiefly, but not
entirely, from the British islands. In the early
days of the Protestant Reformation, there were
European countries in which religious toleration
had advanced practically much further than in
England. The England of Henry VIII. as compared with the Netherlands was in a crude and
backward condition. The contrast might be lik-
ened to that between rural life with its narrow
mental horizon and the varied cosmopolitan life
of the city. England politically was a land of
unrivalled promise, but she was not quite abreast
with the most advanced culture of the time. Her
government was mainly in the hands of country
gentlemen, who lacked some valuable elements of
experience that were possessed by the burghers
of commercial Antwerp and Ghent. A careful

survey of the Middle Ages shows plainly an abiding antagonism between commerce and the ecclesiastical spirit. A general connection between the predominance of international trade and the secularization of public life is distinctly traceable. On the map of mediæval Europe one may point out peculiar spots where the Papacy never gained complete sway. In some of these, as in Bohemia and southern Gaul, the resistance was due to Manichæan heresies brought in from the Eastern Empire, giving rise to a kind of mediæval Puritanism; in these we do not find a spirit of liberal thought developed, but rather an anti-Catholic fanaticism. The other peculiar spots lie in the great pathway of commerce between the Levant and the northern seas. In the free cities of northern Italy and southern Germany, in the Hansa towns, and in the Netherlands, priestcraft had less sway than elsewhere, and the general tone of thought was more liberal and modern. No city came so completely under the secularizing influences of maritime commerce as Venice; and it is significant that the Papacy, at the very pinnacle of its power and arrogance in the thirteenth century, utterly failed in its attempt to force the Inquisition upon that republic of merchants.

In similar wise, we find the commercial Netherlands in the sixteenth century exhibiting practically such toleration in matters of religion as the British islands attained only much later, and after prolonged and distressing struggle. From the time of Edward III. commercial intercourse with the great Dutch and Flemish cities was one of the most potent civilizing influences at work in England. It was a liberalizing influence in religion and in politics, and must be named among the causes which made the eastern counties preëminent for heresy. In later days, when the Dutch provinces had saved their Protestantism and recovered political freedom, they adopted a policy of toleration so broad as to seem to most contemporaries very eccentric. Their noble country was stigmatized as " the common harbour of all heresies " and a " cage of unclean birds." How it harboured heretics escaping from England is something that no American is ever likely to forget.

If, after this glance at European conditions, we cross the Atlantic and observe the group of twelve colonies that were planted during the seventeenth century, we find that five of them were especially notable for pursuing from the outset a policy of toleration, — a policy favourable to liberal thought. These five, naming them in order of seniority, were New Netherland, Maryland, Rhode Island, and

Pennsylvania, with Delaware. In New Netherland the Dutch simply maintained their tradi-
tional secularized policy. On the hospitable island
of Manhattan all the varieties of European religion met on terms of equality, — Lutherans and
Catholics, Quakers [1] and Puritans, Moravians and
Jews. After the English conquest this liberal policy was continued by the bigoted Duke of York,
for reasons similar to those which made toleration
a necessity in the province of the liberal and sagacious Calverts. The Catholic proprietors of Maryland wished to make their province a desirable
home for Catholics who were inclined to leave
England, and the only possible way of accomplish-
ing this, without interference from the British
government, was to pursue a policy broad enough
to include Catholics along with all other kinds of
Christians in its benefits. A similar necessity confronted Charles II. and James II. In order to
secure as much protection as possible for Catholics
without interference from Parliament, it was necessary to pursue a policy broad enough to include
Quakers along with Catholics. For such reasons
James refrained from disturbing the liberal Dutch

---

[1] Stuyvesant's brief persecution of Quakers, for which he was
sternly rebuked by the home government, constitutes an exception to the rule. See my *Dutch and Quaker Colonies*, i. 232-237.

policy in New York. For such reasons both Stuart kings supported the schemes of William Penn, in whose proprietary colonies of Pennsylvania and Delaware the principles of toleration were carried out, on the whole, more completely than anywhere else in English-speaking America. It is interesting in this connection to observe that the mother of William Penn was a Dutch lady, though perhaps it is possible to make too much of such a fact. The Quakers, who formed the strength of the colony, represented a phase of Puritanism. more liberal than Independency. As contrasted with Independency, Quakerism was a notable advance in the direction of Individualism; it had outgrown the set of notions according to which a civic community ought to consist of a united body of believers. Pennsylvania, therefore, and its appendage Delaware, profited by the late date at which they were founded; they represented a more advanced stage of opinion than the colonies which started in the time of James I. Their proprietary government remained undisturbed until the Declaration of Independence, and in 1776 these two states were the only ones in which all Christians, whether Protestant or Catholic, stood socially and politically on an equal footing. For after the accession of William and Mary had made

the Episcopal Church supreme in New York and
Maryland, the Catholic inhabitants of those colonies were disfranchised and made the subject of
various oppressive enactments.   Even the laws of
Rhode Island, as first printed, early in the eighteenth century, expressly prohibit Roman Catholics from voting.   The date of this statute is not
accurately known, but it was certainly between
1688 and 1705,[1] and may be due to the strong
antagonism aroused by the conduct of James II.
and his Jacobites.   However that may be, the
statute was not repealed until 1784.

The disfranchisement of Catholics was contrary
to the spirit of the Rhode Island charter and to
the views of Roger Williams, who certainly understood the rational grounds for religious toleration
better than any other man of his time, save perhaps
Milton and Vane.   He represents the Protestant
principle of the sacred right of private judgment carried out with unflinching logical consist-
ency.   In him the transition from Independency
to Individualism is completed.   The contrast between the two is illustrated in the controversy
between Williams and Cotton which was called
forth by the publication in 1644 of Williams's
book entitled "The Bloudy Tenent of Persecu-

[1] See Arnold's *History of Rhode Island*, ii. 490–496.

tion." John Cotton was a typical Independent, and by no means a man of persecuting temperament, but his view of the matter is extremely one-sided. He admits that it is wrong for error to persecute truth, but he holds it to be the sacred duty of truth to persecute error! Williams, on the other hand, sees that truth stands in no need of violent or artificial support, and that error contains within itself the seeds of death. He feels, too, that when I venture to persecute what I call error in others, I virtually assume my own infallibility. Thus not until pure Individualism is reached is the fundamental fallacy of Catholicism escaped. In order to protect this sacred Individualism, Williams would have a complete separation between church and state. Under no pretext whatever should the civil government interfere with religious matters. There should be no more statutes against heresy or heretics, no enforced attendance upon public worship, no support of churches by taxation. Roger Williams not only proclaimed such doctrines, but he lived up to them. He never took pains to conceal his dislike of Quaker doctrines; in his seventy-third year he once rowed himself in a boat the whole length of Narragansett Bay, in order to conduct a dispute against three valiant Quaker champions; yet, in spite of vehement pressure from the neighbouring colonies, he resolutely refused to allow the civil power of Rhode Island to be used against Quakers. Massachusetts in fury threatened to cut off the trade of the weaker colony, but nothing could intimidate Williams into what he termed " exercising a civil power over men's consciences." Among the public men of the seventeenth century Roger Williams deserves a preëminent place; he was the first to conceive thoroughly and carry out consistently, in the face of strong opposition, a theory of religious liberty broad enough to win assent and approval from advanced thinkers of the present day.

The separation of church from state, which was effected with such remarkable success in the founding of Rhode Island, did not become general in the United States until after the winning of independence. On this issue the eighteenth century had its memorable struggle, in which the protagonist was Virginia, and the victory was achieved under the leadership of Jefferson and Madison. The early policy of Virginia was to drive out dissentients, or subject them to civil disabilities ; and of the Puritans who went thither for a while the greater part left the colony, many of them retreating into tolerant Maryland. After 1660, for three generations

the Episcopal folk had it all their own way.   But
about 1720 began the wholesale immigration of
Presbyterians and Lutherans into the Shenandoah
Valley, and after the middle of the century trouble
began when the tide-water Cavaliers tried to im-
pose taxes upon these people for the support of
the Established Church.   The most numerous and
powerful opponents of this narrow policy were the
Presbyterians; and inasmuch as these had come,
not from Scotland where their own church was established, but from Ireland where it was persecuted,
their experience had led them to approve the
separation of church from state.   Their political
notions were also strongly democratic, and with the
aid of their votes Jefferson's party not only abolished primogeniture and entail and other old English customs, but also carried the disestablishment
of the Episcopal Church in Virginia.   Madison's
Religious Freedom Act of 1785, which not only
effected this, but likewise did away with all religious tests, is a very important event in the history
of the United States.   The statute, which declared
that " opinion in matters of religion shall in no
wise diminish, enlarge, or affect civil capacities,"
attracted attention far and wide ; it was translated
into several European languages, and published
with admiring comments ; and in the course of the

next forty years it was imitated by one state after another, until all over the land religious freedom came to be *almost* as complete as legislation could make it. The qualifying adverb is still needed; for, by the constitutions of Pennsylvania and Tennessee, no man can hold office unless he believes in God and a future state of rewards and punishments; in Texas, Arkansas, Mississippi, the two Carolinas, and Maryland, belief in God is required; and in Arkansas and Maryland a man who does not believe in God and a future state of retribution is deemed incompetent as a witness or juror.[1] Such curiosities of law-making — survivals from a lower state, like the caudal vertebræ in man and the higher apes — are common enough in history.

The various stages here mentioned in the progress toward religious toleration, and toward the separation of church and state, are important symptoms of the progress of liberal thought. Of course Madison's Religious Freedom Act could not have been proposed by an Endicott, or sustained by a community that would not endure the presence of Baptists or Quakers. The sketch here given shows an enormous advance in liberal thought in the course of two centuries and a half. But such a

[1] Stimson, *American Statute Law*, § 46.

survey is far from telling us the whole story. A further inquiry into causal agencies is needed, and the best field for it is furnished by that theocratic Puritanism which cast out Roger Williams, — the Puritanism of the four confederated New England colonies, and especially of Massachusetts. No one can deny that in Massachusetts, during the nineteenth century, liberal thought has advanced further and has permeated the community more thoroughly than in any other state of the American Union. For at least three generations the intellectual ferment upon which liberal thought in the United States has thriven has come chiefly from Massachusetts. Yet among our colonies which attained social maturity during the seventeenth century there was none which made such emphatic exhibitions of intolerance and bigotry as Massachusetts. She was as clearly and avowedly founded upon an illiberal principle as Rhode Island was founded upon a principle of liberality. The Endicott type of mind is the very antipodes of the Roger Williams type ; yet it was in the land of Endicott, and in a congenial soil, that Theodore Parker lately flourished. Whence came so great a change ? The answer will remind us that there are two sources from which liberal thought is nourished. The one is the secularized Gallio spirit that deems it folly to interpose obstacles in the way of the natural working of reason and common sense; the other is the intense devotion to spiritual ideals which, in spite of all inherited encumbrances of bigotry and superstition, never casts off its allegiance to reason as the final arbiter. The former spirit is of vast use in the world, although its tendency is to deaden into mere worldliness as typified in a Franklin; the latter spirit may commit many an error, but its drift is toward light and stimulus and exaltation of life as typified in an Emerson. In the darkest days of New England Puritanism the paramount allegiance to reason was never lost sight of; and out of this fact came the triumph of free thinking, although no such result was ever intended.

The aims of the Puritans who settled in New England were not all alike, but one dominant aim with many was the founding of a commonwealth in which church and state should be identified, somewhat after the pattern of the old Hebrew theocracy. To this end the suffrage in Massachusetts and New Haven was limited to persons qualified to receive the sacrament in Congregational churches. This Massachusetts idea was never adopted by Plymouth, and the founding of Connecticut was at least in part a liberal protest against it. In New Haven it was soon suppressed by the act of Charles II. which put an end to the separate existence of the colony. In Massachusetts, where this theocratic policy prevailed for half a century, the result was the growth of an unenfranchised class which came to include four fifths of the community. During the first generation, when the policy was administered by broadminded, sagacious men like Winthrop and Cotton, its evils were not flagrant. But after 1650, with such fanatics as Norton and the aged Endicott at the helm, it soon became evident that the rulers were at variance on many points with the mass of the people. This was shown with glaring force in the Quaker persecution, when the violence of Endicott's party produced a popular reaction of feeling, which enabled the Quakers to carry their point and remain in the colony in defiance of statutes. It was further shown in the Half-way Covenant and the founding of the Old South Church in 1669, as parts of a movement toward extending the suffrage ; and again in the rise of the Tory party under the lead of Joseph Dudley, opposed to the pretensions of the clergy. The magnificent work of the Massachusetts theocracy in resisting the crown throughout the whole reign of Charles II. can never be forgotten. Nothing

was ever done in America that contributed more
toward the maintenance of political freedom. But
in spite of its merits, the faults of the theocracy
were such that we cannot regret its speedy overthrow. When that overthrow was effected, by the
charter of 1692, there were a great many people
in Massachusetts more or less hostile to the kind
of Puritanism entertained by their grandfathers,
and thus prepared for a more liberal mental habit.
There was also a marked secularization of thought,
a diminution of interest in theological problems,
and a deadening of religious zeal. A wonderful
series of changes was set on foot by the writings
and preaching of Jonathan Edwards, and the
group of revivals between 1735 and 1750 known
as " the Great Awakening." Few figures in history are more pathetic or more sublime than that
of Jonathan Edwards in the lonely woodlands of
Northampton and Stockbridge, a thinker for depth
and acuteness surpassed by not many that have
lived, a man with the soul of poet and prophet,
wrestling with the most terrible problems that
humanity has ever encountered, with more than
the courage and candour of Augustine or Calvin,
with all the lofty inspiration of Fichte or Novalis.
An interesting historical essay might be devoted
to tracing the effects wrought upon New England

by this giant personality. The Great Awakening, in which he took part, and to which his preaching powerfully contributed, revived the popular interest in theological questions, disencumbered of the ever present political implications of the previous century. In many ways his theories acted as a disintegrating solvent upon the beliefs of the time. For example, the prominence which he gave to spiritual conversion, or what was called " change of heart," brought about the overthrow of the doctrine of the Half-way Covenant. It also weakened the logical basis of infant baptism, and led to the winning of hosts of converts by the Baptists. Moreover, the uses to which Edwards put his doctrine of the will produced a reaction toward Arminianism, which not only affected the teachings of the Baptists, but predisposed many persons to join in the wave of Methodism which was just about to sweep over the country. A similar reaction against Edwards's views of divine justice, reinforced by some first faint inklings of Biblical criticism, pointed the way toward Universalism. Still more, the discussions aroused by Edwards's speculations on original sin and the atonement began to undermine the doctrine of the Trinity and prepare men's minds for the Unitarian movement. No such results would have been possible save in a country where education was universal and the Sunday sermon a favourite theme of discussion. Sooner or later, the perpetual appeal to reason, with the familiar use of metaphysical arguments and citations of Scripture, must lead to novelties of doctrine and to negative criticism ; while for the education of the popular intelligence nothing could be more effective. In seventeenth-century Puritanism, therefore, in spite of its rigid narrowness, there were latent the speculations of an Edwards, the further conclusions to which some of them were pushed, the reactions against them, the keen edge of the critical faculty in New England, and much of the free thinking of a later age.

In the course of the eighteenth century some influence was doubtless exercised in America by the English deists, and at the very end of the century by Thomas Paine. There is no reason to suppose that any appreciable effect was produced by the atheism of the French encyclopædists, which was mainly a reaction, largely emotional and aided by the shallowest of metaphysics, against the effete ecclesiastical system in France. It was too remote from American ideas to exert much influence here. The deism of Voltaire found a few scattered admirers. A quiet religion of humanity, which set little store by miracles, or abstruse doctrines,

or the divine authority of Scripture, was held by
a number of eminent persons of strong prosaic
common sense and feeble spirituality, among whom
may be named Franklin and Jefferson and John
Adams. This phase of free thought was of considerable importance, but the dominant influence
in New England down to the rise of the transcendental movement was that which could be traced
back to Edwards.

In the early part of the present century, the
most advanced phase of liberal thought, represented by the Unitarians in Massachusetts, was
trying to hold an utterly untenable position, half-
way between narrow orthodoxy and untrammelled
free thinking, when the ground began to be cut
from under it by the transcendentalists, whose
native temperaments, not wanting in kinship with
that of Edwards, were stimulated by a brief contact with Kantian and post-Kantian speculation in
Germany. In Emerson's poetic soul the result was
a seminal influence upon high thinking, in America
and in the Old World, the power of which we
cannot but feel, but which it is as yet too soon
to estimate. In the middle of the century some
wholesome destructive work still needed to be
done, and it was well done. When German criticism, with the other weapons in the powerful hands

of Theodore Parker, freed us from the spectre of
bibliolatry, it might indeed be said that the promise of the Protestant Reformation was at length
fulfilled. The change wrought in the Unitarian
church since Parker began his preaching has been
to some extent followed by analogous changes in
other churches. On every side, the last quarter of
the nineteenth century has been preëminently the
age of the decomposition of orthodoxies. Here
and there and everywhere they are crumbling into
ruins ; and as the world has long since left behind the age of trilobites and the age of dinosaurs,
so in the world to which we are coming there will
be neither a place nor a use for orthodoxies.

For, as I must observe in conclusion, there is
all about us a resistless and world-wide influence
at work, to which all the temporary and local
causes I have mentioned have been but the ministering servants. From age to age, our knowledge
is growing from more to more. From the discovery of America, from the astronomy of Copernicus
and the physics of Galileo, down to the universal
doctrine of evolution in our own time, there has
been one grand coherent and consecutive tale of
ever enlarging, ever more organized knowledge of
the world in which we live. By this enlarged experience our minds are affected, from day to day

and from year to year, in more ways than we can detect or enumerate. It opens our minds to some notions, and makes them incurably hostile to others; so that, for example, new truths well-nigh beyond comprehension, like some of those connected with the luminiferous ether, are accepted, and old beliefs once universal, like witchcraft, are scornfully rejected. Vast changes in mental attitude are thus wrought before it is generally realized. Into the new scheme of things old beliefs no longer fit, and are therefore thrown aside and forgotten. Now our orthodoxies are of older date than the goodly fabric of modern knowledge. They are the outcome of more primitive and childlike thinking, they have ceased to fit the world as we know it, and therefore they fade and fall away from us, in spite of all our efforts to retain undisturbed the venerable and hallowed associations. In this inevitable struggle there has always been more or less pain, and hence free thought has not usually been popular. It has come to our life feast as a guest unbidden and unwelcome; but it has come to stay with us, and already proves more genial than was expected. Deadening, cramping finality has lost its charm for him who has tasted of the ripe fruit of the tree of knowledge. In this broad universe of God's wisdom and love, not leashes to restrain us are needed, but wings to sustain our flight. Let bold but reverent thought go on and probe creation's mysteries, till faith and knowledge "make one music as before, but vaster."

October, 1895.

VI

SIR HARRY VANE[1]

WITH the single exception of Cromwell, the greatest statesman of the heroic age of Puritanism was unquestionably the younger Henry Vane. He did as much as any one to compass the downfall of Strafford; he brought the military strength of Scotland to the aid of the hard-pressed Parliament; he administered the navy with which Blake won his astonishing victories; he dared even withstand Cromwell at the height of his power, when his measures savoured too much of violence. After the death of Pym in 1643, Sir Henry Vane, then thirty-one years of age, was the foremost man in the Long Parliament, and so remained as long as that Parliament controlled the march of events. As Baxter said, " he was that within the House that Cromwell was without." Yet before the beginning of his brilliant career in England, this

[1] *The Life of Young Sir Henry Vane, Governor of Massachusetts Bay, and Leader of the Long Parliament.* With a Consideration of the English Commonwealth as a Forecast of America. By James K. Hosmer. Boston: Houghton, Mifflin & Co. 1888.

young man had written his name indelibly upon
one of the earliest pages in the history of the
American people. It is pleasant to remember
that this admirable man was once the chief magistrate of an American commonwealth. Thorough
republican and enthusiastic lover of liberty, he was
spiritually akin to Jefferson and to Samuel Adams.
His career furnishes an excellent illustration of Mr.
Doyle's remark, that " by looking at the colony of
Massachusetts, we can see what sort of a commonwealth was constructed by the best men of the Puritan party, and to some extent what they would
have made of the government of England if they
could have had their way unchecked."

An adequate biography of this great statesman
was a thing much to be desired. Half a century
ago Mr. C. W. Upham contributed to Sparks's
" American Biography " an interesting life of
Vane; and about the same time Mr. John Forster,
in his " Statesmen of the Commonwealth," made a
sketch characterized by his usual brilliancy. But
both these writers indulged themselves in that kind
of indiscriminate eulogy which used in those days
to be thought necessary for biographers ; and by
way of foil to their hero they seemed to feel bound
to underrate and misinterpret Cromwell, even as
Carlyle seemed to think he was exalting the great

Protector in belittling Vane. The remarkable advance in fairness and breadth of view which historical studies have made within the last fifty years is nowhere better illustrated than in the spirit in which the seventeenth century in England is treated by Masson and Gardiner as contrasted with Macaulay. It is no longer the fashion to depict individuals or parties as wholly saintlike or quite the reverse, and it is beginning to be practically recognized that there are two sides to almost every question.

The need for an adequate life of Sir Harry Vane has been most thoroughly and admirably satisfied by Mr. Hosmer. As a biography and as a historical monograph, it deserves to be ranked among the best books of the day. It paints a lifelike picture of the man, and it describes, in a broad, generous spirit and with keen philosophical insight, the causal succession of events in one of the most momentous political contests the world has ever seen. We are getting far enough away from the seventeenth century to realize the critical importance of the struggle in which kingship was struck down in England just as it was attaining unchecked supremacy in all the other great nations of Europe. We can put the Great Rebellion into its proper place in the series of conflicts which have so far

resulted in spreading constitutional government
far and wide over two hemispheres; and we can
begin to see how disastrous in its consequences
would have been the victory of the Cavaliers, true
and gallant men as most of them doubtless were.
Without dealing too much in generalities, Mr.
Hosmer's narrative keeps before us the gravity
of the issues at stake, while our attention is seldom drawn away from the powerful but quiet and
gracious personality that occupies the centre of the
canvas. It is customary for great eras to live in
the twilight of popular memory in association with
some one surpassing name, while other heroes of
the time are dimly remembered or quite forgotten.
The work of these other men gets unconsciously
transferred to the credit of the most brilliant or
striking hero, as Hamilton, for example, is apt
to get associated not merely with his own all-im-
portant achievements, but likewise with those of
Madison and the Federal Convention generally.
In accordance with this labour-saving habit of
mind, the Great Rebellion in popular memory
means Oliver Cromwell, while such men as Eliot
and Pym, Fairfax and Ireton, are passed over; and
if Hampden stays, it is partly due to the often-
quoted line of the poet Gray. So there are many
who know Vane only through Milton's sonnet, —

itself perhaps the noblest literary tribute ever paid
to a statesman. In Mr. Hosmer's pages Sir Harry
lives again, one of the brightest figures of the Puritan age, cheerful and affectionate, full of sacred
enthusiasm, yet shrewd and self-contained. " He
was indeed a man of extraordinary parts, a pleasant wit, a great understanding which pierced into
and discerned the purposes of men with wonderful
sagacity, whilst he had himself *vultum clausum,*
that no man could make a guess of what he intended." So says Clarendon, who loved him not,
but could not help admiring the skill which, at the
most critical moment of the war, when many stout
adherents of the parliamentary cause were inclined
to abandon it as lost, all at once brought light out
of darkness, as the signing of the Solemn League
and Covenant summoned Alexander Leslie and
twenty thousand brawny Scots across the border
to stand side by side with Cromwell and Fairfax
at Marston Moor. In later days it became matter
of common report that the northern Covenanters
had fallen a prey to the wiles of " that sweet
youth," and allowed themselves to be hoodwinked
and cozened by " sly Sir Harry," until, in the hope
of establishing Presbyterianism south of the Tweed,
they lent themselves to the work of setting the
monster Independency upon its feet. Mr. Hosmer

carefully examines this charge, and, we think, successfully refutes it. It was neither the first nor the last contract on record which has afterward come to receive conflicting interpretations from the two parties without any tricksome intent on either side. "The Scots," says Mr. Hosmer, "understood that England assumed their own narrow Presbyterianism, with its complete intolerance; Vane and his friends gave the instrument a different interpretation, which they honestly felt it would bear." The amendments which Vane partly succeeded in engrafting upon the Scottish proposals at Edinburgh are sufficient evidence of his straightforwardness. It was plain enough that, in making a league to overcome the King, the Scots wanted one thing, while the English wanted another. Vane did not hide this fact; to have emphasized it would have been to forfeit all claim to diplomatic tact. His part in the memorable negotiation is tersely summed up by Clarendon: "Sir Harry Vane was one of the commissioners, and therefore the others need not be named, since he was all in any business where others were joined with him." In the Committee of Both Kingdoms which the league created he was equally effective, and it was mainly through his persistent dexterity that the committee acquired the control of military affairs, and thus gave to the operations of the parliamentary army that unity which they had hitherto lacked.

The firstfruits of Vane's diplomacy were Marston Moor and Naseby, and it would be unreasonable to find fault with Mr. Hosmer for pausing to describe those battles. They are brilliant episodes in his narrative. We have nowhere seen the two battles more lucidly explained. The author has been himself a soldier, and has looked at the ground with a military eye. One quite envies him the pleasant journey, as on his tricycle he follows the route of the Ironsides over the smooth roads and smiling fields of Merry England. His pages are redolent of the mellow cheer and fragrance of the summer day under that mild northern sun. One catches, with the author, the spirit of the deadly fight, and realizes, as Naseby spire fades away in the distance, the gravity of the crisis and the completeness of the victory. Said stout old Sir Jacob Astley, when the Roundheads took him captive a few months afterward, " Gentlemen, ye may now sit down and play, for you have done all your work, if you fall not out among yourselves."

They were already falling out among themselves; how seriously, Dunbar and Worcester were by and by to show. "Their own generation," says Mr. Hosmer, "believed that the Independents drew their origin from America." Certainly there had been witnessed in Boston, in the year when Harvard College was founded, some noteworthy manifestations of Independency, and scenes had been enacted which had left a deep impress upon Sir Harry's youthful mind. In 1635 the gossips wrote: "Sir Henry Vane hath as good as lost his eldest son, who is gone into New England for conscience' sake; he likes not the discipline of the Church of England; . . . no persuasions of our bishops nor authority of his parents could prevail with him: let him go." The fascinating boy arrived in Boston in October, 1635, and in the following March, having won all hearts, was elected governor of Massachusetts. He witnessed the Pequot war, the beautiful heroism and rare diplomacy of Roger Williams, and the bitter strife which ensued upon the teachings of Mrs. Hutchinson. Mr. Hosmer gives a vivid picture of the life in the little colony, the theological warfare, and the passionate tears of the young man as the difficulties thickened around him. Perhaps his indiscreet threat of an appeal to the throne in favour of the Antinomians, as he sailed for England in the summer of 1637, may have served to hasten the

banishment of Mrs. Hutchinson ; but the lesson
of toleration was already taking shape in his mind,
as was clearly shown in his controversy with Winthrop. His friendly relations with Roger Williams
began at the time of the Pequot war ; and in 1643,
when Williams visited England in quest of a charter for Rhode Island, he was Vane's guest at his
house in London, and also at his country seat in
Lincolnshire. It was then that Williams wrote
that noble book, " The Bloudy Tenent of Persecution for Cause of Conscience," in the preface to
which he thus refers to his friend : " Mine ears
were glad and late witnesses of an heavenly speech
of one of the most eminent of that High Assembly
of Parliament : *Why should the labours of any
be suppressed, if sober, though never so different?
We now profess to seek God, we desire to see
light !* " [1] Mr. Hosmer gives in facsimile a touching letter from Vane to Winthrop in 1645, in which
he urges his friends in New England to respect the
liberty of conscience.

In 1648, in order to save the cause of liberty
from losing to intrigue and chicanery all the
ground it had won by the sword, the Ironsides
felt themselves called upon to take things into
their own hands. This period of the story, ex-

[1] See my *Beginnings of New England*, p. 185.

tending to the forcible dissolution of the Rump
Parliament in 1653, Mr. Hosmer treats under the
rubric of American England. For the moment,
the spirit of Independency, which reigned supreme
in Massachusetts, asserted itself in England in the
temporary overthrow of the crown and the aristo-
cracy. In this period Sir Harry appears as the
opponent of the extreme measures of his party.
He heartily disapproves of such irregular proceedings as Pride's Purge and the execution of the
King. Here is shown the strong conservatism of
temperament of this law-abiding American-English-
man. He had all the ingrained reverence of our
sturdy practical race for constitutional methods,
and withal a far-sighted intelligence that could
discern ways of settling the difficulty which were
for the moment impracticable, because his contemporaries had not grown up to them. In his mind
were the rudiments of the idea of a written constitution, upon which a new government for England
might be built, with powers neatly defined and
limited. One fancies that in some respects he
would have felt himself more at home if he could
have been suddenly translated from the Rump
Parliament of 1653 to the Federal Convention
of 1787, in which immortal assembly there sat
perhaps no man of loftier spirit than his. It

was natural enough that Cromwell, whose stern
common sense discerned the practical need of the
moment and reluctantly fulfilled it, should cry,
" The Lord deliver me from Sir Harry Vane ! "
In spite of this antagonism at the supreme crisis,
however, the Protector recognized the worth of his
opponent, and seems to have borne him no deep-
seated ill will. There was no downright break
between them until the Healing Question came up,
in 1656.

In Vane's last years there seemed to be some
good reasons for distrusting his judgment on practical questions. The element of dreamy enthusiasm always present in him began to come into the
foreground as his more sober ideas and plans were
thwarted. Some of his latest utterances are like
the rhapsodies of the Fifth Monarchists. Herein
again appears his spiritual kinship with his friends
in Massachusetts. The theocratic ideal of the
founders of Massachusetts, as developed freely in
the American wilderness, was kept within rational
bounds ; but if hemmed in by such inexorable circumstances as checked the early growth of repub-
licanism in England, it would very likely have
flowered grotesquely enough in Fifth Monarchist
vagaries. From Edward Johnson, of Woburn,
author of the " Wonder-Working Providence,"

there often came the dithyrambic utterances of an extreme Fifth Monarchy man.

When Charles II. came back to his father's throne, there was but one thing to be done with such a representative republican as Sir Harry Vane. His head must come off, for there was not room enough in England to hold him and the son of Charles I. at the same time. He died on Tower Hill, with all the fearlessness and charming sweetness that had always marked his life. His memory is a precious possession for all coming generations ; and the book in which Mr. Hosmer has told the story of his life, with such warm sympathy and such broad intelligence, is worthy of its subject.

January, 1889.

VII

THE ARBITRATION TREATY

AFTER negotiations which had been pending for nearly two years, the general Arbitration Treaty between the United States and Great Britain was signed on the 11th of January [1897] by Mr. Richard Olney and Sir Julian Pauncefote, representing the two countries concerned; and on the following day the document was sent by President Cleveland to the Senate for ratification. The provisions of this important treaty may be summarized as follows : —

It is expected that differences arising between the two countries will ordinarily admit of settlement by the customary methods of diplomacy. It is only with cases where such customary methods fail that the provisions of the present treaty are concerned; and the parties hereby agree to submit all such cases to arbitration after the manner herein provided.

The "questions in difference" that are liable to arise are arranged in three grades or classes: (1) small pecuniary claims; (2) large pecuniary claims, and others not involving questions of territory; (3) territorial claims. For each of these grades there is to be a special method of settlement.

First, " all pecuniary claims or groups of claims, which in the aggregate do not exceed $500,000 in amount and do not involve the determination of territorial claims," shall be decided by a tribunal constituted as follows: " Each party shall nominate one arbitrator, who shall be a jurist of repute, and the two arbitrators so nominated shall, within two months of their nomination, select an umpire. In the event of their failing to do so within the limit of time, the umpire shall be appointed by agreement between the members of the Supreme Court of the United States and the members of the Judicial Committee of the Privy Council in Great Britain." In case these persons fail to agree upon an umpire within three months, the King of Sweden and Norway shall appoint one. Among public personages of unquestionable dignity and importance, this sovereign is as likely as any to be free from bias against either the United States or Great Britain; but should either party object to him, they may adopt a substitute, if they can agree upon one. It does not seem likely that the failure to select an umpire would often reach the stage where an appeal to the Swedish King would be necessary.

The umpire, when and however appointed, shall be president of the tribunal of three, and the award of a majority of the members shall be final. Under these provisions, it may be expected that all petty claims can be disposed of without unreasonable delay, and with as little risk of unfairness as one would find in any court whatever.

Secondly, "all pecuniary claims or groups of claims exceeding $500,000, and all other matters in respect whereof either of the parties shall have rights against the other, under the treaty or otherwise, provided they do not involve territorial claims," shall be dealt with as follows: Such claims must be submitted to the tribunal of three, as above described, and its award, if unanimous, shall be final. If the award is not unanimous, either party may demand a review of it, but such demand must be made within six months from the date of the award. In such case, the appellate tribunal shall consist of five jurists of repute, no one of whom has been a member of the tribunal of three whose award is to be reviewed. Of these five jurists, two shall be selected by each party, and these four shall agree upon their umpire within three months after their nomination. In case of their failure, the umpire shall be selected (as in the former case) by the members of the Supreme

Court and the Judicial Committee of the Privy
Council; and if these do not agree within three
months, the selection shall be left (as before) to the
King of Sweden and Norway. The umpire, when
selected, shall preside. The award of the tribunal
of three shall be reviewed by this tribunal of five,
and the award of a majority of the five shall be
final.

Thirdly, "any controversy involving the determination of territorial claims shall be submitted to
a tribunal of six members," three of whom shall be
judges of the Supreme Court or of Circuit Courts,
to be nominated by the President of the United
States. The other three shall be members of the
highest British court or members of the Judicial
Committee of the Privy Council, to be nominated
by the Queen. "Their award by a majority of
not less than five to one shall be final. If there is
less than the prescribed majority, the award shall
also be final, unless either party within three
months protests that the award is erroneous. If
the award is protested, or if the members of
the tribunal are equally divided, there shall be
no recourse to hostile measures of any description until the mediation of one or more friendly
powers shall have been invited by one or the other
party." It is also provided that "where one of

the United States or a British colony is specially
concerned, the President or Queen may make a judicial officer of the state or colony an arbitrator."

In some cases, a question may be removed from
the jurisdiction of the tribunal of three or the
tribunal of five, and transferred to that of the
tribunal of six. If, prior to the close of the hearing of the claim before the lower tribunal, it shall
be decided by the tribunal, upon the motion of
either party, that the determination of the claim
necessarily involves a decision of some "disputed
question of principle of grave general importance,
affecting the national rights of such party as distinct from its private rights, of which it is merely
an international representative," then the jurisdiction of the lower tribunal over the claim shall at
once cease, and it shall be dealt with by the tribunal of six.

With regard to territorial claims, a special article defines them as including not only all claims to
territory, but also "all other claims involving questions of servitude, rights of navigation, access to
fisheries, and all rights and interests necessary to
control the enjoyment of either's territory."

The treaty is to remain in force for five years
from the date at which it becomes operative, and
"until a year after either party shall have notified
the other of its wish to terminate it."

The first impression which one gets from reading the treaty is that it is strictly defined and limited in its application. Yet, when duly considered, it seems to cover all chances of controversy that are likely to arise between the United States and Great Britain. Under such a treaty as this, nearly all the questions at issue between the two countries since 1783 might have been satisfactorily adjusted, — the payment of private debts to British creditors, the relinquishment of the frontier posts by British garrisons, the northeastern boundary, the partition of the Oregon territory, the questions concerning the Newfoundland fisheries, the navigation of the Great Lakes, the catching of seals in Bering Sea, the difference of opinion over the San Juan boundary, etc. Possibly some of the old questions growing out of the African slave trade might have been brought within its purview, but that is now of small consequence, since no issues of that sort are likely ever to rise again. Differences attending the future construction of a Nicaragua canal, regarded as an easement or a servitude possibly affecting vested rights, might, under a liberal interpretation, be dealt with; and one may suppose that the Venezuela question is meant to be covered, since it relates to territorial claims in which, though they may not obviously concern the

United States either immediately or remotely, our government has with unexpected emphasis declared itself interested.

On the other hand, one does not seem to find in the treaty any provision which would have covered two or three of the most serious questions that have ever been in dispute between the United States and Great Britain. One of these questions, concerning the right of search and the impressment of seamen, was conspicuous among the causes of the ill-considered and deplorable War of 1812. But it may be presumed, with strong probability, that no difficulty of that kind can again arise between these two powers. The affair of the Trent in 1861 seems also to be a kind of case not provided for. But that affair, most creditably settled at a moment of fierce irritation and under aggravating circumstances, was settled in such wise as to establish a great principle which will make it extremely difficult for such a case to occur again. As for the Alabama Claims, they could apparently have been adjusted under the present treaty, as large pecuniary claims involving international principles of grave general importance.

On the whole, there seems to be small likelihood of any dispute arising between this country and Great Britain which cannot be amicably settled, with reasonable promptness, under the provisions of this new Arbitration Treaty. One chief desideratum in any such instrument is to secure impartiality in the arbitrating tribunals, and here the arrangements made in our treaty will doubtless yield as good results as can ever be achieved through mere arrangements. In such matters, the best of machinery is of less consequence than the human nature by which the machinery is to be worked. Impartiality, not only real, but conspicuous and unmistakable, is the prime requisite in a court of arbitration. Its life and health can be sustained only in an atmosphere of untainted and unsuspected integrity. But in an age which does not yet fully comprehend the damnable villainy of such maxims as " Our country, right or wrong," gross partisanship is not easy to eliminate from human nature. Even austere judges, taken from a Supreme Court, have sometimes shown themselves to be men of like passions with ourselves. It would need but few awards made on the " eight to seven " principle, as in the Electoral Commission of 1877, to make our arbitrating tribunal the laughing-stock of the world, and to set back for a generation or two the hand upon the timepiece of civilization.

A general experience, however, justifies us in hoping much better things from the group of international tribunals contemplated in our present treaty. There is no doubt that the good work is undertaken in entire good faith by both nations; both earnestly wish to make international arbitration successful, and there is little fear that the importance of fair dealing will be overlooked or undervalued. If the present proceedings result in the establishment of a tribunal whose integrity and impartiality shall win the permanent confidence of British and Americans alike, it will be an immense achievement, fraught with incalculable benefit to mankind. For the first time, the substitution of international lawsuits for warfare will have been systematically begun by two of the leading nations of the world; and an event which admits of such a description cannot be without many consequences, enduring and profound.

For observe that the interest of the present treaty lies not so much in the fact that it provides for arbitration as in the fact that it aims at making arbitration the regular and permanent method of settling international disputes. In due proportion to the gravity of the problem is the modest caution with which it is approached. The treaty merely asks to be tried on its merits, and only for five years at that. Only for such a brief period is the most vociferous Jingo in the United States

Senate or elsewhere asked to put a curb upon his sanguinary propensities and see what will happen. Nay, if we really prefer war to peace; if, like the giant in the nursery tale, we are thirsting for a draught of British blood, neither this nor any other treaty could long restrain us. As Hosea Biglow truly observes,—

> "The right to be a cussed fool
> Is safe from all devices human."

It has been rumoured that some Senators will vote against the treaty, in order to show their spite against President Cleveland and Mr. Olney. If the treaty should fail of confirmation through such a cause, it would be no more than has happened before. Members of the Sapsea family have sat in other chambers than those of the Capitol at Washington. But, as a rule, good causes have not long been hindered through such pettiness, and should the treaty thus fail for the moment, it would not be ruined, but only delayed. In any event, it is not likely to be long in acquiring its five years' lease of life. If during that time nothing should occur to discredit it, even should no cases arise to call it into operation, its purpose is so much in harmony with the most enlightened spirit of the age that it is pretty sure to be renewed. Should cases arise under it, the machinery which it provides

is confessedly provisional and tentative, and upon
renewal can be modified in such wise as may seem
desirable. Other human institutions have been
moulded by experience, and so, doubtless, it will
be with international courts of arbitration.

The working of the tribunals created by the present treaty will be carefully watched by other nations than the two parties directly concerned, and
should it achieve any notable success it will furnish
a precedent likely to be imitated. The removal of
any source of irritation at all comparable to the
Alabama Claims would be, of course, a success of
the first magnitude; great good, with far-reaching
consequences, might be wrought by a much smaller
one. Probably few readers are aware of the extent to which the arbitration at Geneva in 1872
has already served as a precedent for the peaceful
solution of international difficulties.[1]　Already the

[1] The following list of instances within a period of twelve years
is cited from an able article by Professor Pasquale Fiore, of the
University of Naples, in the *International Journal of Ethics*, October, 1896 : —

Arbitration by the Emperor of Austria between Great Britain
and Nicaragua, 1881.

A mixed commission to arbitrate between France and Chili, 1882.

Arbitration by the President of the French Republic between
the Netherlands and the Republic of San Domingo, 1882.

Arbitration by Pope Leo XIII. between Germany and Spain ;
affair of the Caroline Islands, 1885.

moral effect of that event has been such as to suggest that it may hereafter be commemorated as the illustrious herald of a new era. The Geneva event was brought about by a treaty specially framed for the purpose, and might thus be regarded as exceptional or extraordinary in its nature. Still greater, then, would be the moral effect of a similar success achieved by a tribunal created under the provisions of a permanent treaty.

The commission to arbitrate between the Argentine Republic and Brazil, 1886.

Arbitration by Spain between Colombia and Venezuela, 1887.

Arbitration by the minister of Spain at Bogotá between Italy and Colombia, 1887.

Arbitration by President Cleveland between Nicaragua and Costa Rica, 1888.

Arbitration by the Queen of Spain between Peru and Ecuador, 1888.

Arbitration by Baron Lambermont between England and Germany; affair of Lamoo, 1888.

Arbitration by the Czar of Russia between France and the Netherlands; affair of the boundaries of Guinea, 1888.

Arbitration by Sir Edward Momson between Denmark and Sweden, 1888.

Compromise between the United States and Venezuela, 1890.

Compromise between Germany, the United States, and Great Britain; affair of Terranova, 1891.

Arbitration by Switzerland between England, the United States, and Portugal; affair of the railroads at Delagoa Bay, 1891.

Arbitration between Great Britain and the United States relating to the question of the delimitation of territorial power in Bering Sea, 1893.

It may be urged that arbitration cannot often
succeed in dealing with difficulties so formidable
as those connected with the Alabama Claims. The
questions hitherto settled by arbitration have for
the most part been of minor importance, in which
" national honour " has not been at stake, and the
bestial impulse to tear and bruise, which so many
light-headed persons mistake for patriotism, has
not been aroused. The London "Spectator" tells
us that if the United States should ever repeat the
Mason and Slidell incident, or should feel insulted
by the speech of some British prime minister, there
would be war, no matter how loudly the lawyers
in both countries might appeal to the Arbitration
Treaty. The two illustrations cited are not happy
ones, since from both may be deduced reasons
why war is not likely to ensue. The Mason and
Slidell incident was a most impressive illustration
of the value of delay and discussion in calming
popular excitement. The principle of international law which the United States violated on
that occasion was a principle for which the United
States had long and earnestly contended against
the opposition of Great Britain. A very brief discussion of the affair in the American press made
this clear to every one, and there was no cavilling
when our government disowned the act and sur-

rendered the prisoners with the noble frankness which characterized President Lincoln's way of doing things. What chiefly tended to hinder or prevent such a happy termination of the affair was the unnecessary arrogance of Lord Palmerston's government in making its demand of us. What chiefly favoured it was the absence of an ocean telegraph, affording the delay needful for sober second thought. I remember hearing people say at the time that the breaking of the first Atlantic cable in 1858 had thus turned out to be a blessing in disguise! Now, should any incident as irritating as the Trent affair occur in future, the Arbitration Treaty can be made to furnish the delay which the absence of an ocean cable once necessitated ; and I have enough respect for English-speaking people on both sides of the water to believe that in such case they will behave sensibly, and not like silly duellists. So, too, as regards "feeling insulted" by the speech of a prime minister, there is a recent historic instance to the point. Our British cousins may have had reason to feel insulted by some expressions in President Cleveland's message of December, 1895, but they took the matter very quietly. Had the boot been on the other leg, a few pupils of Elijah Pogram might have indulged in Barmecide suppers of gore, but there the affair

would probably have ended. The reason is that
deliberate public opinion in both countries feels
sure that nothing is to be gained, and much is to
be lost, by fighting. Under such conditions, the
growing moral sentiment which condemns most
warfare as wicked has a chance to assert itself.
Thus the delay which allows deliberate public
opinion to be brought to bear upon irritating
incidents is a great advantage ; and the mere
existence of a permanent arbitration treaty tends
toward insuring such delay.

People who prefer civilized and gentleman-like
methods of settling disputes to the savage and
ruffian-like business of burning and slaughtering
are sometimes stigmatized by silly writers as " sen-
timentalists." In the deliberate public opinion
which has come to be so strong a force in preventing war between the United States and Great
Britain, sentiment has as yet probably no great
place ; but it is hoped and believed that it will by
and by have much more. In the days of Alexander Hamilton, there was very little love for the
Federal Union in any part of this country ; it was
accepted as a disagreeable necessity. But his policy
brought into existence a powerful group of selfish
interests binding men more and more closely to
the Union, and more so at the North than at the

South. When Webster made his reply to Hayne, there was a growing sentiment of Union for him to appeal to, and stronger at the North than at the South. When the Civil War came, that sentiment was strong enough to sadden the heart of many a Southerner whose sense of duty made him a secessionist; at the North it had waxed so powerful that men were ready to die for it, as the Mussulman for his Prophet or the Cavalier for his King. Thus sentiment can quickly and sturdily grow when favoured by habits of thought originally dictated by self-interest. Obviously a state of things in favour of which a strong sentiment is once enlisted has its chances of permanence greatly increased. I therefore hope and believe that in the deliberate public opinion above mentioned sentiment will by and by have a larger place than it has at present. As feelings of dislike between the peoples of two countries are always unintelligent and churlish, so feelings of friendship are sure to be broadening and refining. The abiding sentiment of Scotchmen toward England was for many centuries immeasurably more rancorous than any Yankee schoolboy ever gave vent to on the Fourth of July. There is no reason why the advent of the twenty-first century should not find the friendship between the United States and Great Britain quite as strong as that between Scotland and England to-day. Toward so desirable a consummation a permanent policy of arbitration must surely tend.

The fact that deliberate public opinion in both countries can be counted upon as strongly adverse to war is the principal fact which makes such a permanent policy feasible. It is our only sufficient guarantee that the awards of the international tribunal will be respected. These considerations need to be borne in mind, if we try to speculate upon the probable influence upon other nations of a successful system of arbitration between the United States and Great Britain. Upon the continent of Europe a considerable interest seems already to have been felt in the treaty, and, as I observed above, its working is sure to be carefully watched ; for the states of Europe are suffering acutely from the apparent necessity of keeping perpetually prepared for war, and any expedient that holds out the slightest chance of relief from such a burden cannot fail to attract earnest attention.

The peoples of Europe are not unfamiliar with the principles of arbitration. Indeed, like many other good things which have loomed up conspicuously in recent times, arbitration can be traced back to the ancient Greeks, for whom it occasionally mitigated the evils attendant upon frequent warfare between their city - states. Among the Italian republics of the Middle Ages, disputes were sometimes submitted to the arbitration of learned professors in the universities at Bologna and other towns. But such methods could not prevail over the ruder fashions of Europe north of the Alps. As mediæval Italy was the industrial and commercial centre of the world, so in our day it is the nations most completely devoted to industry and commerce, the English-speaking nations, that are foremost in bringing into practice the methods of arbitration. The settlement of the Alabama Claims is the most brilliant instance on record, and we have already cited examples of the readiness of sundry nations, great and small, to imitate it. Such examples, even when concerned with questions of minor importance, are to some extent an indication of the growing conviction that war, and the unceasing preparations for it, are becoming insupportable burdens.

It is the steadily increasing complication of industrial life, and the heightened standard of living that has come therewith, that are making men, year by year, more unwilling to endure the burdens entailed by war. In the Middle Ages, human life was made hideous by famine, pestilence, perennial warfare, and such bloody superstitions as the belief in witchcraft ; but men contrived to endure it, because they had no experience of anything better, and could not even form a conception of relief save such as the Church afforded. Deluges of war, fraught with horrors which stagger our powers of conception, swept at brief intervals over every part of the continent of Europe, and the intervals were mostly filled with petty waspish raids that brought robbery and murder home to everybody's door ; while honest industry, penned up within walled towns, was glad of such precarious immunity as stout battlements eked out by blackmail could be made to afford. Fighting was incessant and ubiquitous. The change wrought in six centuries has been amazing, and it has been chiefly due to industrial development. Private warfare has been extinguished, famine and pestilence seldom occur in civilized countries, mental habits nurtured by science have banished the witches, the land is covered with cheerful homesteads, and the achievement of success in life through devotion to industrial pursuits has become general. Wars have greatly diminished in frequency, in length, and in the amount of misery needlessly inflicted. We have thus learned how pleasant life can become under peaceful conditions, and we are determined as far as possible to prolong such conditions. We have no notion of submitting to misery like that of the Middle Ages; on the contrary, we have got rid of so much of it that we mean to go on and get rid of the whole. Such is the general feeling among civilized men. It may safely be said not only that no nation in Christendom wishes to go to war, but also that the nations are few which would not make a considerable sacrifice of interests and feelings rather than incur its calamities. For reasons such as these, the states of Continental Europe are showing an increasing disposition to submit questions to arbitration, and in view of this situation the fullest measure of success for our Arbitration Treaty is to be desired, for the sake of its moral effect.

The method at present in vogue on the continent of Europe for averting warfare is the excessively cumbrous expedient of keeping up great armaments in time of peace. The origin of this expedient may be traced back to the *levée en masse* to which revolutionary France resorted in the agonies of self-defence in 1792. The *levée en masse* proved to be a far more formidable engine of warfare than the small standing armies with which Europe had long been familiar; and so, after the old military system of Prussia had been overthrown in 1806,

the reforms of Stein and Scharnhorst introduced
the principle of the *levée en masse* into times of
peace, dividing the male population into classes
which could be kept in training, and might be successively called to the field as soon as military exigencies should demand it. The prodigious strength
which Prussia could put forth under this system was
revealed in 1866 and 1870, and since then similar
methods have become universally adopted, so that
the commencement of a general European war to-
day would doubtless find several millions of men
under arms. The progress of invention is at the
same time daily improving projectiles on the one
hand, and fortifications on the other; we may perhaps hope that some of us will live long enough
to see what will happen when a ball is fired with
irresistible momentum against an impenetrable
wall! To keep up with the progress of invention
enormous sums are expended on military engines,
while each nation endeavours to avert war by making such a show of strength as will deter other
nations from attacking it. A mania for increasing
armaments has thus been produced, and although
this state of things is far less destructive and de-
moralizing than actual war, it lays a burden upon
Europe which is fast becoming intolerable. For
the modern development of industry has given rise

to problems that press for solution, and no satisfactory solution can be reached in the midst of this monstrous armed peace. Competition has reached a point where no nation can afford to divert a considerable percentage of its population from industrial pursuits. Each nation, in order to maintain its rank in the world, is called upon to devote its utmost energies to agriculture, manufactures, and commerce. Moreover, the economic disturbances due to the withdrawal of so many men from the work of production are closely connected with the discontent which finds vent in the wild schemes of socialists, communists, and anarchists. There is no other way of beginning the work of social redemption but by a general disarmament; and this opinion has for some years been gaining strength in Europe. It is commonly felt that in one way or another the state of armed peace will have to be abandoned.

In a lecture at the Royal Institution of Great Britain in 1880, I argued that the contrast between the United States, with a population quite freed from the demands of militarism, and the continent of Europe, with its enormous armaments useless for productive purposes, could not long be maintained; that American competition would soon come to press so severely upon Europe as to compel a disarmament, and in this way the swords would get beaten into ploughshares. American competition is less effective than it might be, owing to our absurd tariffs and vicious currency, but its tendency has undoubtedly been in the direction indicated. I suspect, however, that the process will be less simple. Within the last twenty years the operations of production and distribution have been ' assuming colossal proportions. Syndicates, trusts, and other huge combinations of capital have begun carrying on business upon a scale heretofore unprecedented. Already we see symptoms that such combinations are to include partners in various parts of the earth. Business, in short, is becoming more and more international; and under such circumstances the era of general disarmament is likely to be hastened. In the long run, peace has no other friend so powerful as commerce.

While every successful resort to arbitration is to be welcomed as a step toward facilitating disarmament, it seems probable that institutions of somewhat broader scope than courts of arbitration will be required for the settlement of many complex international questions. In the European congresses which have assembled from time to time to deal with peculiar exigencies, we have the precedent for such more regular and permanent institutions. An example of what is meant was furnished by the Congress of Paris in 1856, when it dealt summarily with the whole group of vexed questions relating to the rights and duties of neutrals and belligerents upon the ocean, and put an end to the chaos of two centuries by establishing an international code relating to piracy, blockades, and seizures in times of naval war. This code has been respected by maritime powers and enforced by the world's public opinion, and its establishment was a memorable incident in the advance of civilization. Now, such work as the Congress of Paris did can be done in future by other congresses, but it is work of broader scope than has hitherto been undertaken by courts of arbitration. I am inclined to think that both these institutions — the International Congress and the Tribunal of Arbitration — are destined to survive, with very considerable increase in power and dignity, in the political society of the future, long after disarmament has become an accomplished fact.

About the time that a small party of Englishmen at Jamestown were laying the first foundation stones of the United States, one of the greatest kings and one of the greatest ministers of modern times were deeply engaged in what they called the Great Design, a scheme for a European Confedera-

tion.   The plan of Henry IV. of France and the
Duke of Sully contemplated a federal republic
of Christendom, comprising six hereditary crowns
(France, England, Spain, Sweden, Denmark, Lom-
bardy), five elective crowns (the Empire, the Papacy, Bohemia, Hungary, Poland), and four republics (Venice, the small Italian states, Switzerland, and the Netherlands).   There was to be a
federal government in three branches, legislative,
executive, judicial ; a federal army of about three
hundred thousand men, and a powerful federal
fleet.   The purpose of the federation was to put
an end once and forever to wars, both civil and
international.   Probably the two great statesmen
were not sanguine as to the immediate success of
their Great Design, and doubtless none knew better than they that it would cost at least one mighty
war to establish it.   But there is a largeness of
view about the scheme that is refreshing to meet in
a world of arid and narrow commonplaces.   With
all their breadth of vision, however, Henry and
Sully would surely have been amazed had they
been told that the handful of half-starved Englishmen at Jamestown were inaugurating a political
and social development that in course of time
would contribute powerfully toward the success of
something like their Great Design.

In human affairs a period of three centuries is a brief one, and the progress already made in the direction toward which the two great Frenchmen were looking is significant and prophetic. The vast armaments now maintained on the continent of Europe cannot possibly endure. Economic necessities will put an end to them before many years. But disarmament, apparently, can only proceed *pari passu* with the establishment of peaceful methods of settling international questions. The machinery for this will probably be found in the further development of two institutions that have already come into existence, the International Congress and the Court of Arbitration. The existence of these institutions, which is now occasional, will tend to become permanent : the former will deal preferably with the establishment of general principles, the latter with their judicial application to special cases. As European congresses meet now upon extraordinary occasions, so once it was with the congresses of the American colonies, such as the New York Congress of 1690 and the Albany Congress of 1754 for concerting measures against New France, and the New York Congress of 1765 for protesting against the Stamp Act. Then came the Continental Congress of 1774, which circumstances kept in existence for fifteen years, until a

political revolution reached its consummation in
replacing it by a completely organized federal government.   In 1754 the possibility of a permanent
federation of American states was derided as an
idle dream of Benjamin Franklin and Thomas
Hutchinson.   Very little love was lost between
the people of different colonies; and when the
crisis came on, after 1783, the majority hated and
dreaded a permanent Federal Union, and accepted
it *only as the alternative to something worse*,
namely, anarchy and civil war.   In like manner,
it may be surmised as not improbable that in
course of time the occasions for summoning European congresses will recur with increasing frequency until the functions which they are called
upon to discharge will convert them into a permanent institution.   Such a development, combined
with the increased employment of arbitration, must
ultimately tend toward the creation of a Federal
Union in Europe.   The fact that such a result
will be hated and dreaded by many people, perhaps by the great majority, need not prevent its
being accepted and acquiesced in *as the alternative
to something worse*, namely, the indefinite contin-
uance of the system of vast armaments.

By the time when such a result comes clearly
within sight, it will very likely have been made

evident that the policy of isolation which our country has wisely pursued for the century past cannot be maintained perpetually. When Washington wrote his Farewell Address, the danger of our getting dragged into the mighty struggle then raging in Europe was a real and serious danger, against which we needed to be solemnly warned. Since then times have changed, and they are changing still. From a nation scarcely stronger than Portugal we have become equal to the strongest. Railways, telegraphs, and international industries are making every part of the world the neighbour of every other part. To preserve a policy of isolation will not always be possible, nor will it be desirable. Situations will arise (if they have not already arisen) in which such moral weight as the United States can exert will be called for. The pacification of Europe, therefore, is not an affair that is foreign to our interests. In that, as in every other aspect of the Christian policy of " peace on earth and good will to men," we are most deeply concerned ; and every incident, like the present Arbitration Treaty, that promises to advance us even by one step toward the sublime result, it is our solemn duty to welcome and encourage by all the means within our power.

February, 1897.

VIII

FRANCIS PARKMAN[1]

In the summer of 1865 I had occasion almost daily to pass by the pleasant windows of Little, Brown & Co., in Boston, and it was not an easy thing to do without stopping for a moment to look in upon their ample treasures. Among the freshest novelties there displayed were to be seen Lord Derby's translation of the Iliad, Forsyth's Life of Cicero, Colonel Higginson's Epictetus, a new edition of Edmund Burke's writings, and the tasteful reprint of Froude's History of England, just in from the Riverside Press. One day, in the midst of such time-honoured classics and new books on well-worn themes, there appeared a stranger that claimed attention and aroused curiosity. It was a modest crown octavo, clad in sombre garb, and bearing the title " Pioneers of France in the

[1] This paper originated in an address at Sanders Theatre, Cambridge, December 6, 1893, at a service commemorative of Mr. Parkman. In its present greatly expanded shape it was printed as the Introduction to the revised edition of Parkman's Works, Boston, 1897–98, 20 vols., octavo.

New World." The author's name was not familiar to me, but presently I remembered having seen
it upon a stouter volume labelled "The Conspiracy
of Pontiac," of which many copies used to stand
in a row far back in the inner and dusky regions
of the shop. This older book I had once taken
down from its shelf, just to quiet a lazy doubt as
to whether Pontiac might be the name of a man
or a place. Had that conspiracy been an event in
Merovingian Gaul or in Borgia's Italy, I should
have felt a twinge of conscience at not knowing
about it; but the deeds of feathered and painted
red men on the Great Lakes and the Alleghanies,
only a century old, seemed remote and trivial. Indeed, with the old-fashioned study of the humani-
ties, which tended to keep the Mediterranean too
exclusively in the centre of one's field of vision, it
was not always easy to get one's historical perspective correctly adjusted. Scenes and events that
come within the direct line of our spiritual ancestry, which until yesterday was all in the Old
World, thus become unduly magnified, so as to
deaden our sense of the interest and importance of
the things that have happened since our forefathers
went forth from their homesteads to grapple with
the terrors of an outlying wilderness. We find no
difficulty in realizing the historic significance of

Marathon and Chalons, of the barons at Runny-
mede or Luther at Wittenberg; and scarcely a
hill or a meadow in the Roman's Europe but
blooms for us with flowers of romance. Literature and philosophy, art and song, have expended
their richest treasures in adding to the witchery
of Old World spots and Old World themes.

But as we learn to broaden our horizon, the
perspective becomes somewhat shifted. It begins
to dawn upon us that in New World events, also,
there is a rare and potent fascination. Not only
is there the interest of their present importance,
which nobody would be likely to deny, but there
is the charm of a historic past as full of romance
as any chapter whatever in the annals of mankind.
The Alleghanies as well as the Apennines have
looked down upon great causes lost and won, and
the Mohawk Valley is classic ground no less than
the banks of the Rhine. To appreciate these
things thirty years ago required the vision of a
master in the field of history; and when I carried
home and read the "Pioneers of France," I saw
at once that in Francis Parkman we had found
such a master. The reading of the book was for
me, as doubtless for many others, a pioneer experience in this New World. It was a delightful
experience, repeated and prolonged for many a

year, as those glorious volumes came one **after another** from the press, until the story of the struggle
between France and England for the possession of
North America was at last completed. It was an
experience of which the full significance required
study in many and apparently diverse fields to
realize. By step after step one would alight upon
new ways of regarding America and its place in
universal history.

First and most obvious, plainly visible from the
threshold of the subject, was its extreme picturesqueness. It is a widespread notion that American history is commonplace and dull; and as for
the American red man, he is often thought to
be finally disposed of when we have stigmatized
him as a bloodthirsty demon and grovelling beast.
It is safe to say that those who entertain such
notions have never read Mr. Parkman. In the
theme which occupied him his poet's eye saw nothing that was dull or commonplace. To bring
him vividly before us, I will quote his own words
from one of the introductory pages of his opening
volume : —

" The French dominion is a memory of the past;
and when we evoke its departed shades, they rise
upon us from their graves in strange romantic
guise. Again their ghostly camp fires seem to

burn, and the fitful light is cast around on lord and
vassal and black-robed priest, mingled with wild
forms of savage warriors, knit in close fellowship
on the same stern errand. A boundless vision
grows upon us: an untamed continent; vast wastes
of forest verdure; mountains silent in primeval
sleep; river, lake, and glimmering pool; wilderness oceans mingling with the sky. Such was the
domain which France conquered for civilization.
Plumed helmets gleamed in the shade of its for-
ests, priestly vestments in its dens and fastnesses
of ancient barbarism. Men steeped in antique
learning, pale with the close breath of the cloister,
here spent the noon and evening of their lives,
ruled savage hordes with a mild parental sway,
and stood serene before the direst shapes of death.
Men of courtly nurture, heirs to the polish of a
far-reaching ancestry, here with their dauntless
hardihood put to shame the boldest sons of toil."

When a writer in sentences that are mere generalizations gives us such pictures as these, one
has much to expect from his detailed narrative,
glowing with sympathy and crowded with incident.
In Parkman's books such expectations are never
disappointed. What was an uncouth and howling
wilderness in the world of literature he has taken
for his own domain, and peopled it forever with

living figures, dainty and winsome, or grim and
terrible, or sprightly and gay. Never shall be
forgotten the beautiful earnestness, the devout serenity, the blithe courage, of Champlain; never can
we forget the saintly Marie de l'Incarnation, the
delicate and long-suffering Lalemant, the lionlike
Brébeuf, the chivalrous Maisonneuve, the grim
and wily Pontiac, or that man against whom fate
sickened of contending, the mighty and masterful
La Salle. These, with many a comrade and foe,
have now their place in literature as permanent
and sure as Tancred or St. Boniface, as the Cid
or Robert Bruce. As the wand of Scott revealed
unsuspected depths of human interest in Border
castle and Highland glen, so it seems that North
America was but awaiting the magician's touch
that should invest its rivers and hillsides with
memories of great days gone by. Parkman's sweep
has been a wide one, and many are the spots that
his wand has touched, from the cliffs of the Sague-
nay to the Texas coast, and from Acadia to the
western slopes of the Rocky Mountains.

I do not forget that earlier writers than Parkman had felt something of the picturesqueness and
the elements of dramatic force in the history of
the conquest of our continent. In particular, the
characteristics of the red men and the incidents of

forest life had long ago been made the theme of
novels and poems, such as they were; I wonder
how many people of to-day remember even the
names of such books as "Yonnondio" or "Ka-
baosa"? All such work was thrown into the shade
by that of Fenimore Cooper, whose genius, though
limited, was undeniable. But when we mention
Cooper we are brought at once by contrast to the
secret of Parkman's power. It has long been
recognized that Cooper's Indians are more or less
unreal; just such creatures never existed anywhere. When Corneille and Racine put ancient
Greeks or Romans on the stage they dressed them
in velvet and gold lace, flowing wigs and high
buckled shoes, and made them talk like Louis
XIV.'s courtiers; in seventeenth-century dramatists the historical sense was lacking. In the next
age it was not much better. When Rousseau had
occasion to philosophize about men in a state of
nature he invented the Noble Savage, an insuffer-
able creature whom any real savage would justly
loathe and despise. The noble savage has figured
extensively in modern literature, and has left his
mark upon Cooper's pleasant pages as well as
upon many a chapter of serious history. But you
cannot introduce unreal Indians as factors in the
development of a narrative without throwing a

shimmer of unreality about the whole story. It is like bringing in ghosts or goblins among live men and women : it instantly converts sober narrative into fairy tale ; the two worlds will no more mix than oil and water. The ancient and mediæval minds did not find it so, as the numberless histories encumbered with the supernatural testify ; but the modern mind does find it so. The modern mind has taken a little draught, the prelude to deeper draughts, at the healing and purifying well of science ; and it has begun to be dissatisfied with anything short of exact truth. When any unsound element enters into a narrative, the taint is quickly tasted, and its flavour spoils the whole.

We are then brought, I say, to the secret of Parkman's power. His Indians are true to the life. In his pages Pontiac is a man of warm flesh and blood, as much so as Montcalm or Israel Putnam. This solid reality in the Indians makes the whole work real and convincing. Here is the great contrast between Parkman's work and that of Prescott, in so far as the latter dealt with American themes. In reading Prescott's account of the conquest of Mexico, one feels one's self in the world of the " Arabian Nights ; " indeed, the author himself, in occasional comments, lets us see that he is unable to get rid of just such a feeling.

His story moves on in a region that is unreal to him, and therefore tantalizing to the reader; his Montezuma is a personality like none that ever existed beneath the moon. This is because Prescott simply followed his Spanish authorities not only in their statements of physical fact, but in their inevitable misconceptions of the strange Aztec society which they encountered; the Aztecs in his story are unreal, and this false note vitiates it all. In his Peruvian story Prescott followed safer leaders in Garcilasso de la Vega and Cieza de Leon, and made a much truer picture; but he lacked the ethnological knowledge needful for coming into touch with that ancient society, and one often feels this as the weak spot in a narrative of marvellous power and beauty.

Now it was Parkman's good fortune at an early age to realize that in order to do his work it was first of all necessary to know the Indian by personal fellowship and contact. It was also his good fortune that the right sort of Indians were still accessible. What would not Prescott have given, what would not any student of human evolution give, for a chance to pass a week or even a day in such a community as the Tlascala of Xicotencatl

or the Mexico of Montezuma! That phase of social development has long since disappeared. But
fifty years ago, on our great western plains and
among the Rocky Mountains, there still prevailed
a state of society essentially similar to that which
greeted the eyes of Champlain upon the St. Lawrence and of John Smith upon the Chickahominy.
In those days the Oregon Trail had changed but
little since the memorable journey of Lewis and
Clark in the beginning of the present century.
In 1846, two years after taking his bachelor degree at Harvard, young Parkman had a taste
of the excitements of savage life in that primeval
wilderness. He was accompanied by his kinsman,
Mr. Quincy Shaw. They joined a roving tribe of
Sioux Indians, at a time when to do such a thing
was to take their lives in their hands, and they
spent a wild summer among the Black Hills of
Dakota and in the vast moorland solitudes through
which the Platte River winds its interminable
length. In the chase and in the wigwam, in
watching the sorcery of which their religion chiefly
consisted, or in listening to primitive folk tales by
the evening camp fire, Parkman learned to understand the red man, to interpret his motives and his
moods. With his naturalist's keen and accurate
eye and his quick poetic apprehension, that youth-

ful experience formed a safe foundation for all his future work. From that time forth he was fitted to absorb the records and memorials of the early explorers, and to make their strange experiences his own.

The next step was to gather these early records from government archives, and from libraries public and private, on both sides of the Atlantic, — a task, as Parkman himself called it, " abundantly irksome and laborious." It extended over many years and involved several visits to Europe. It was performed with a thoroughness approaching finality. Already in the preface to the " Pioneers " the author was able to say that he had gained access to all the published materials in existence. Of his research among manuscript sources a notable monument exists in a cabinet now standing in the library of the Massachusetts Historical Society, containing nearly two hundred folio volumes of documents copied from the originals by expert copyists. Ability to incur heavy expense is, of course, a prerequisite for all undertakings of this sort, and herein our historian was favoured by fortune. Against this chiefest among advantages were to be offset the hardships entailed by delicate health and inability to use the eyes for reading and writing. Parkman always dic-

tated instead of holding the pen, and his huge
mass of documents had to be read aloud to him.
The heroism shown year after year in contending
with physical ailments was the index of a character
fit to be mated, for its pertinacious courage, with
the heroes that live in those shining pages.

The progress in working up materials was slow
and sure. "The Conspiracy of Pontiac," which
forms the sequel and conclusion of Parkman's
work, was first published in 1851, only five years
after the summer spent with the Indians; fourteen years then elapsed before the "Pioneers"
made its appearance in Little, Brown & Co.'s
window; and then there were yet seven - and-
twenty years more before the final volumes came
out in 1892. Altogether, about half a century
was required for the building of this grand literary monument. Nowhere can we find a better
illustration of the French critic's definition of a
great life, — a thought conceived in youth, and
realized in later years.

This elaborateness of preparation had its share
in producing the intense vividness of Parkman's
descriptions. Profusion of detail makes them seem
like the accounts of an eye-witness. The realism
is so strong that the author seems to have come
in person fresh from the scenes he describes, with

the smoke of the battle hovering about him and its
fierce light glowing in his eyes. Such realism is
usually the prerogative of the novelist rather than
of the historian, and in one of his prefaces Parkman recognizes that the reader may feel this and
suspect him. " If at times," he says, " it may
seem that range has been allowed to fancy, it is
so in appearance only, since the minutest details
of narrative or description rest on authentic documents or on personal observation."

This kind of personal observation Parkman carried so far as to visit all the important localities,
indeed well-nigh all the localities, that form the
scenery of his story, and study them with the patience
of a surveyor and the discerning eye of a landscape
painter. His strong love of nature added keen
zest to this sort of work. From boyhood he was a
trapper and hunter; in later years he became eminent as a horticulturist, originating new varieties
of flowers. To sleep under the open sky was his
delight. His books fairly reek with the fragrance
of pine woods. I open one of them at random, and
my eye falls upon such a sentence as this: " There
is softness in the mellow air, the warm sunshine,
and the budding leaves of spring; and in the forest
flower, which, more delicate than the pampered offspring of gardens, lifts its tender head through the

refuse and decay of the wilderness." Looking at the context, I find that this sentence comes in a remarkable passage suggested by Colonel Henry Bouquet's western expedition of 1764, when he compelled the Indians to set free so many French and English prisoners. Some of these captives were unwilling to leave the society of the red men ; some positively refused to accept the boon of what was called freedom. In this strange conduct, exclaims Parkman, there was no unaccountable perversity ; and he breaks out with two pages of noble dithyrambics in praise of savage life. " To him who has once tasted the reckless independence, the haughty self-reliance, the sense of irresponsible freedom, which the forest life engenders, civilization thenceforth seems flat and stale. . . . The entrapped wanderer grows fierce and restless, and pants for breathing room. His path, it is true, was choked with difficulties, but his body and soul were hardened to meet them ; it was beset with dangers, but these were the very spice of his life, gladdening his heart with exulting self-confidence, and sending the blood through his veins with a livelier current. The wilderness, rough, harsh, and inexorable, has charms more potent in their seductive influence than all the lures of luxury and sloth. And often he on whom it has cast its

magic finds no heart to dissolve the spell, and remains a wanderer and an Ishmaelite to the hour
of his death." [1]

No one can doubt that the man who could write
like this had the kind of temperament that could
look into the Indian's mind and portray him correctly. But for this inborn temperament all his
microscopic industry would have availed him but
little. To use his own words: "Faithfulness to
the truth of history involves far more than a
research, however patient and scrupulous, into special facts. Such facts may be detailed with the
most minute exactness, and yet the narrative, taken
as a whole, may be unmeaning or untrue." These
are golden words for the student of the historical
art to ponder. To make a truthful record of a
vanished age patient scholarship is needed, and
something more. Into the making of a historian
there should enter something of the philosopher,
something of the naturalist, something of the poet.
In Parkman this rare union of qualities was realized in a greater degree than in any other American historian. Indeed, I doubt if the nineteenth
century can show in any part of the world another
historian quite his equal in respect of such a union.

There is one thing which lends to Parkman's

[1] *Pontiac*, iii. 112.

work a peculiar interest, and will be sure to make
it grow in fame with the ages. Not only has he
left the truthful record of a vanished age so complete and final that the work will never need to be
done again, but if any one should in future attempt
to do it again he cannot approach the task with
quite such equipment as Parkman. In an important sense, the age of Pontiac is far more remote
from us than the age of Clovis or the age of Aga-
memnon. When barbaric society is overwhelmed
by advancing waves of civilization, its vanishing is
final; the thread of tradition is cut off forever
with the shears of Fate. Where are Montezuma's
Aztecs? Their physical offspring still dwell on
the table-land of Mexico, and their ancient speech
is still heard in the streets, but that old society is
as extinct as the trilobites, and has to be painfully
studied in fossil fragments of custom and tradition.
So with the red men of the North : it is not true
that they are dying out physically, as many people
suppose, but their stage of society is fast disappearing, and soon it will have vanished forever. Soon
their race will be swallowed up and forgotten, just
as we overlook and ignore to-day the existence
of five thousand Iroquois farmers in the state of
New York.

Now the study of comparative ethnology has

begun to teach us that the red Indian is one of the most interesting of men. He represents a stage of evolution through which civilized men have once passed, — a stage far more ancient and primitive than that which is depicted in the Odyssey or in the Book of Genesis. When Champlain and Frontenac met the feathered chieftains of the St. Lawrence, they talked with men of the Stone Age face·to face. Phases of life that had vanished from Europe long before Rome was built survived in America long enough to be seen and studied by modern men. Behind Mr. Parkman's picturesqueness, therefore, there lies a significance far more profound than one at first would suspect. He has portrayed for us a wondrous and forever fascinating stage in the evolution of humanity. We may well thank Heaven for sending us such a scholar, such an artist, such a genius, before it was too late. As we look at the changes wrought in the last fifty years, we realize that already the opportunities by which he profited in youth are in large measure lost. He came not a moment too soon to catch the fleeting light and fix it upon his immortal canvas.

Thus Parkman is to be regarded as first of all the historian of Primitive Society. No other great historian has dealt intelligently and consecutively

with such phases of barbarism as he describes with
such loving minuteness. To the older historians
all races of men very far below the European
grade of culture seemed alike ; all were ignorantly
grouped together as " savages." Mr. Lewis Morgan first showed the wide difference between true
savages, such as the Apaches and Bannocks on the
one hand, and barbarians with developed village
life, like the Five Nations and the Cherokees. The
latter tribes in the seventeenth and eighteenth
centuries exhibited social phenomena such as were
probably witnessed about the shores of the Mediterranean some seven or eight thousand years earlier. If we carry our thoughts back to the time
that saw the building of the Great Pyramid, and
imagine civilized Egypt looking northward and
eastward upon tribes of white men with social
and political ideas not much more advanced than
those of Frontenac's red men, our picture will be
in its most essential features a correct one. What
would we not give for a historian who, with a
pen like that of Herodotus, could bring before us
the scenes of that primeval Greek world before
the cyclopean works at Tiryns were built, when the
ancestors of Solon and Aristides did not yet dwell
in neatly joinered houses and fasten their door-
latches with a thong, when the sacred city-state

was still unknown, and the countryman had not
yet become a bucolic or "tender of cows," and
butter and cheese were still in the future! No
written records can ever take us back to that time
in that place; for there, as everywhere in the
eastern hemisphere, the art of writing came many
years later than the domestication of animals, and
some ages later than the first building of towns.
But in spite of the lack of written records, the
comparative study of institutions, especially comparative jurisprudence, throws back upon those prehistoric times a light that is often dim, but sometimes wonderfully suggestive and instructive. It
is a light that reveals among primeval Greeks ideas
and customs essentially similar to those of the
Iroquois. It is a light that grows steadier and
brighter as it leads us to the conclusion that five
or six thousand years before Christ white men
around the Ægean Sea had advanced about as far
as the red men in the Mohawk Valley two centuries ago. The one phase of this primitive society
illuminates the other, though extreme caution is
necessary in drawing our inferences. Now Parkman's minute and vivid description of primitive
society among red men is full of lessons that may
be applied with profit to the study of preclassic
antiquity in the Old World. No other historian

has brought us into such close and familiar contact with human life in such ancient stages of its progress. In Parkman's great book we have a record of vanished conditions such as hardly exists anywhere else in literature.

I say his great book, using the singular number; for, with the exception of that breezy bit of autobiography, " The Oregon Trail," all Parkman's books are the closely related volumes of a single comprehensive work. From the adventures of " The Pioneers of France " a consecutive story is developed through " The Jesuits in North America" and " The Discovery of the Great West." In " The Old Régime in Canada" it is continued with a masterly analysis of French methods of colonization in this their greatest colony, and then from " Frontenac and New France under Louis XIV." we are led through " A Half-Century of Conflict " to the grand climax in the volumes on " Montcalm and Wolfe," after which " The Conspiracy of Pontiac " brings the long narrative to a noble and brilliant close. In the first volume we see the men of the Stone Age at that brief moment when they were disposed to adore the bearded newcomers as Children of the Sun; in the last we read the bloody story of their last and most desperate concerted effort to loosen the iron grasp with which these palefaces had seized and were holding the continent. It is a well-rounded tale, and as complete as anything in real history, where completeness and finality are things unknown.

Between the beginning and the end of this well-rounded tale a mighty drama is wrought out in all its scenes. The struggle between France and England for the soil of North America was one of the great critical moments in the career of mankind, — no less important than the struggle between Greece and Persia, or between Rome and Carthage. Out of the long and complicated interaction between Roman and Teutonic institutions which made up the history of the Middle Ages, two strongly contrasted forms of political society had grown up and acquired aggressive strength when in the course of the sixteenth century a New World beyond the sea was laid open for colonization. The maritime nations of Europe were naturally the ones to be attracted to this new arena of enterprise ; and Spain, Portugal, France, England, and Holland each played its interesting and characteristic part. Spain at first claimed the whole, excepting only that Brazilian coast which Borgia's decree gave to Portugal. But Spain's methods, as well as her early failure of strength, prevented her from making good her claim. Spain's methods were limited

to stepping into the place formerly occupied by the
conquering races of half-civilized Indians. She
made aboriginal tribes work for her, just as the
Aztec Confederacy and the Inca dynasty had done.
Where she was brought into direct contact with
American barbarism without the intermediation of
half-civilized native races, she made little or no
headway. Her early failure of strength, on the
other hand, was due to her total absorption in the
fight against civil and religious liberty in Europe.
The failure became apparent as soon as the absorption had begun to be complete. Spain's last
aggressive effort in the New World was the destruction of the little Huguenot colony in Florida in
1565, and it is at that point that Parkman's great
work appropriately begins. From that moment
Spain simply beat her strength to pieces against
the rocks of Netherland courage and resourceful-
ness. As for the Netherlands, their energies were
so far absorbed in taking over and managing the
great Eastern empire of the Portuguese that their
work in the New World was confined to seizing
upon the most imperial geographical position, and
planting a cosmopolitan colony there that, in the
absence of adequate support, was sure to fall into
the hands of one or the other of the competitors
more actively engaged upon the scene.

The two competitors thus more actively engaged were **France and England,** and from an early period it was felt between the two to be a combat in which no quarter was to be given or accepted. These two strongly contrasted forms of political society had each its distinct ideal, and that ideal was to be made to prevail, to the utter exclusion and destruction of the other. Probably the French perceived this somewhat earlier than the English; they felt it to be necessary to stamp out the English before the latter had more than realized the necessity of defending themselves against the French. For the type of political society represented by **Louis XIV.** was preëminently militant, as the English type was preëminently industrial. The aggressiveness of the former was more distinctly conscious of its own narrower aims, and was more deliberately set at work to attain them, while the English, on the other hand, rather drifted into a tremendous world fight without distinct consciousness of their purpose. Yet after the final issue had been joined, the refrain *Carthago delenda est* was heard from the English side, and it came fraught with impending doom from the lips of Pitt as in days of old from the lips of Cato.

The French idea, had it prevailed in the strife, would not have been capable of building up a pacific union of partially independent states, covering this vast continent from ocean to ocean. Within that rigid and rigorous bureaucratic system there was no room for spontaneous individuality, no room for local self-government, and no chance for a flexible federalism to grow up. A well-known phrase of Louis XIV. was, " The state is myself." That phrase represented his ideal. It was approximately true in Old France, realized as far as sundry adverse conditions would allow. The Grand Monarch intended that in New France it should be absolutely true. Upon that fresh soil was to be built up a pure monarchy without concession to human weaknesses and limitations. It was a pet scheme of Louis XIV., and never did a philanthropic world-mender contemplate his grotesque phalanstery or pantarchy with greater pleasure than this master of kingcraft looked forward to the construction of a perfect Christian state in America.

The pages of our great historian are full of examples which prove that if the French idea failed of realization, and the state it founded was overwhelmed, it was not from any lack of lofty qualities in individual Frenchmen. In all the history of the American continent no names stand higher than some of the French names. For courage, for fortitude and high resolve, for sagacious leadership, statesmanlike wisdom, unswerving integrity, devoted loyalty, for all the qualities which make life heroic, we may learn lessons innumerable from the noble Frenchmen who throng in Parkman's pages. The difficulty was not in the individuals, but in the system; not in the units, but in the way they were put together. For while it is true — though many people do not know it — that by no imaginable artifice can you make a society that is better than the human units you put into it, it is also true that nothing is easier than to make a society that is worse than its units. So it was with the colony of New France.

Nowhere can we find a description of despotic government more careful and thoughtful, or more graphic and lifelike, than Parkman has given us in his volume on "The Old Régime in Canada." Seldom, too, will one find a book fuller of political wisdom. The author never preaches like Carlyle, nor does he hurl huge generalizations at our heads like Buckle; he simply describes a state of society that has been. But I hardly need say that his description is not — like the Dryasdust descriptions we are sometimes asked to accept as history — a mere mass of pigments flung at random upon a canvas. It is a picture painted with consummate

art; and in this instance the art consists in so
handling the relations of cause and effect as to
make them speak for themselves. These pages
are alive with political philosophy, and teem with
object lessons of extraordinary value. It would be
hard to point to any book where History more
fully discharges her high function of gathering
friendly lessons of caution from the errors of the
past.

Of all the societies that have been composed of
European men, probably none was ever so despot-
ically organized as New France, unless it may have
been the later Byzantine Empire, which it resem-
bled in the minuteness of elaborate supervision over
all the pettiest details of life. In Canada the pro-
tective, paternal, socialistic, or nationalistic theory
of government — it is the same old cloven hoof,
under whatever specious name you introduce it —
was more fully carried into operation than in any
other community known to history except ancient
Peru. No room was left for individual initiative
or enterprise. All undertakings were nationalized.
Government looked after every man's interests in
this world and the next : baptized and schooled
him ; married him and paid the bride's dowry ; gave
him a bounty with every child that was born to
him ; stocked his cupboard with garden seeds and

compelled him to plant them; prescribed the size
of his house and the number of horses and cattle he
might keep, and the exact percentages of profit
he might be allowed to make, and how his chim-
neys should be swept, and how many servants he
might employ, and what theological doctrine he
might believe, and what sort of bread the bakers
might bake, and where goods might be bought and
how much might be paid for them; and if in a society so well cared for it were possible to find indi-
gent persons, such paupers were duly relieved, from
a fund established by government. Unmitigated
benevolence was the theory of Louis XIV.'s Canadian colony, and heartless political economy had
no place there. Nor was there any room for free
thinkers; when the King after 1685 sent out word
that no mercy must be shown to heretics, the governor, Denonville, with a pious ejaculation, replied
that not so much as a single heretic could be found
in all Canada.

Such was the community whose career our historian has delineated with perfect soundness of judgment and wealth of knowledge. The fate of this
nationalistic experiment, set on foot by one of the
most absolute of monarchs and fostered by one of
the most devoted and powerful of religious organizations, is traced to the operation of causes in-

herent in its very nature. The hopeless paralysis,
the woeful corruption, the moral torpor, resulting
from the suppression of individualism, are vividly
portrayed ; yet there is no discursive generalizing,
and from moment to moment the development of
the story proceeds from within itself. It is the
whole national life of New France that is displayed
before us. Historians of ordinary calibre exhibit
their subject in fragments, or they show us some
phases of life and neglect others. Some have no
eyes save for events that are startling, such as battles and sieges ; or decorative, such as coronations
and court balls. Others give abundant details of
manners and customs ; others have their attention
absorbed by economics ; others again feel such interest in the history of ideas as to lose sight of mere
material incidents. Parkman, on the other hand,
conceives and presents his subject as a whole. He
forgets nothing, overlooks nothing ; but whether
it is a bloody battle, or a theological pamphlet, or
an exploring journey though the forest, or a code
for the discipline of nunneries, each event grows
out of its context as a feature in the total development that is going on before our eyes. It is only
the historian who is also philosopher and artist that
can thus deal in block with the great and complex
life of a whole society. The requisite combination

is realized only in certain rare and high types of mind, and there has been no more brilliant illustration of it than Parkman's volumes afford.

The struggle between the machine-like socialistic despotism of New France and the free and spontaneous political vitality of New England is one of the most instructive object lessons with which the experience of mankind has furnished us. The depth of its significance is equalled by the vastness of its consequences. Never did Destiny preside over a more fateful contest; for it determined which kind of political seed should be sown all over the widest and richest political garden plot left untilled in the world. Free industrial England pitted against despotic militant France for the possession of an ancient continent reserved for this decisive struggle, and dragging into the conflict the belated barbarism of the Stone Age,— such is the wonderful theme which Parkman has treated. When the vividly contrasted modern ideas and personages are set off against the romantic though lurid background of Indian life, the artistic effect becomes simply magnificent. Never has historian grappled with another such epic theme, save when Herodotus told the story of Greece and Persia, or when Gibbon's pages resounded with the solemn tread of marshalled hosts through a thousand years of change.

The story of Mr. Parkman's life can be briefly told. He was born in Boston, in what is now known as Allston Street, September 16, 1823. His ancestors had for several generations been honourably known in Massachusetts. His great-grandfather, Rev. Ebenezer Parkman, a graduate of Harvard in 1741, was minister of the Congregational church in Westborough for nearly sixty years; he was a man of learning and eloquence, whose attention was not all given to Calvinistic theology, for he devoted much of it to the study of history. A son of this clergyman, at the age of seventeen, served as private in a Massachusetts regiment in that greatest of modern wars which was decided on the Heights of Abraham. How little did this gallant youth dream of the glory that was by and by to be shed on the scenes and characters passing before his eyes by the genius of one of his own race and name! Another son of Ebenezer Parkman returned to Boston and became a successful merchant, engaged in that foreign traffic which played so important and liberalizing a part in American life in the days before the Enemy of mankind had invented forty per cent tariffs. The home of this merchant, Samuel Parkman, on the corner of Green and Chardon streets, was long famous for its beautiful flower garden,

indicating perhaps the kind of taste and skill so
conspicuous afterwards in his grandson.   In Samuel the clerical profession skipped one generation,
to be taken up again by his son, Rev. Francis
Parkman, a graduate of Harvard in 1807, and for
many years after 1813 the eminent and beloved
pastor of the New North Church.   Dr. Parkman
was noted for his public spirit and benevolence.
Bishop Huntington, who knew him well, says of
him: "Every aspect of suffering touched him ten-
derly.   There was no hard spot in his breast.
His house was the centre of countless mercies to
various forms of want; and there were few soli-
citors of alms, local or itinerant, and whether for
private necessity or public benefactions, that his
doors did not welcome and send away satisfied.
. . . For many years he was widely known and es-
teemed for his efficient interest in some of our most
conspicuous and useful institutions of philanthropy.
Among these may be especially mentioned the
Massachusetts Bible Society, the Society for Pro-
pagating the Gospel, the Orphan Asylum, the
Humane Society, the Medical Dispensary, the Society for the Relief of Aged and Destitute Clergy-
men, and the Congregational Charitable Society."
He also took an active interest in Harvard University, of which he was an Overseer.   In 1829

he founded there the professorship of " Pulpit Eloquence and the Pastoral Care," familiarly known
as the Parkman Professorship.   A pupil and friend
of Channing, he was noted among Unitarians for
a broadly tolerant disposition.   His wealth of practical wisdom was enlivened by touches of mirth,
so that it was said that you could not " meet
Dr. Parkman in the street, and stop a minute to
exchange words with him, without carrying away
with you some phrase or turn of thought so exquisite in its mingled sagacity and humour that it
touched the inmost sense of the ludicrous, and
made the heart smile as well as the lips."   Such
was the father of our historian.

Mr. Parkman's mother was a descendant of
Rev. John Cotton, one of the most eminent of the
leaders in the great Puritan exodus of the seventeenth century.   She was the daughter of Nathaniel
Hall, of Medford, member of a family which was
represented in the convention that framed the Constitution of Massachusetts in 1780.   Caroline Hall
was a lady of remarkable character, and many of her
fine qualities were noticeable in her distinguished
son.   Of her the late Octavius Frothingham says:
" Humility, charity, truthfulness, were her prime
characteristics.   Her conscience was firm and lofty,
though never austere.   She had a strong sense of

right, coupled with perfect charity toward other people; inflexible in principle, she was gentle in practice. Intellectually she could hardly be called brilliant or accomplished, but she had a strong vein of common sense and practical wisdom, great penetration into character, and a good deal of quiet humour."

Of her six children, the historian, Francis Parkman, was the eldest. As a boy his health was delicate. In a fragment of autobiography, written in the third person, he tells us that "his childhood was neither healthful nor buoyant," and "his boyhood, though for a time active, was not robust." There was a nervous irritability and impulsiveness which kept driving him into activity more intense than his physical strength was well able to bear. At the same time an inborn instinct of self-control, accompanied, doubtless, by a refined unwillingness to intrude his personal feelings upon the notice of other people, led him into such habits of self-repression that his friends sometimes felicitated him on "having no nerves." There was something rudely stoical in his discipline. As he says: "It was impossible that conditions of the nervous system abnormal as his had been from infancy should be without their effects on the mind, and some of these were of a nature highly to exasperate him.

Unconscious of their character and origin, and ignorant that with time and confirmed health they would have disappeared, he had no other thought than that of crushing them by force, and accordingly applied himself to the work. Hence resulted a state of mental tension, habitual for several years, and abundantly mischievous in its effects. With a mind overstrained and a body overtasked, he was burning his candle at both ends."

The conditions which were provided for the sensitive and highly strung boy during a part of his childhood were surely very delightful, and there can be little doubt that they served to determine his career. His grandfather Hall's home in Medford was situated on the border of the Middlesex Fells, a rough and rocky woodland, four thousand acres in extent, as wild and savage in many places as any primeval forest. The place is within eight miles of Boston, and it may be doubted if anywhere else can be found another such magnificent piece of wilderness so near to a great city. It needs only a stray Indian or two, with a few bears and wolves, to bring back for us the days when Winthrop's company landed on the shores of the neighbouring bay. In the heart of this shaggy woodland is Spot Pond, a lake of glorious beauty, with a surface of three hundred acres, and a homely name which it is to be hoped it may always keep, — a name bestowed in the good old times before the national vice of magniloquence had begun to deface our maps. Among the pleasure drives in the neighbourhood of Boston, the drive around Spot Pond is perhaps foremost in beauty. A few fine houses have been built upon its borders, and well-kept roads have given to some parts of the forest the aspect of a park, but the greater part of the territory is undisturbed, and will probably remain so. Seventy years ago the pruning hand of civilization had scarcely touched it. To his grandfather's farm, on the outskirts of this enchanting spot, the boy Parkman was sent in his eighth year. There, he tells us, " I walked twice a day to a school of high but undeserved reputation, about a mile distant, in the town of Medford. Here I learned very little, and spent the intervals of schooling more profitably in collecting eggs, insects, and reptiles, trapping squirrels and woodchucks, and making persistent though rarely fortunate attempts to kill birds with arrows. After four years of this rustication I was brought back to Boston, when I was unhappily seized with a mania for experiments in chemistry, involving a lonely, confined, unwholesome sort of life, baneful to body and mind." No doubt the experience of

four years of plastic boyhood in Middlesex Fells
gave to Parkman's mind the bent which directed
him toward the history of the wilderness.   This
fact he recognized of himself in after life, while he
recalled those boyish days as the brightest in his
memory.

At the age of fifteen or so the retorts and crucibles were thrown away forever, and a reaction in
favor of woodland life began ; "a fancy," he says,
"which soon gained full control over the course of
the literary pursuits to which he was also addicted."
Here we come upon the first mention of the combination of interests which determined his career.
A million boys might be turned loose in Middlesex
Fells, one after another, there to roam in solitude until our globe should have entered upon a new geological period, and the chances are against any one of
them becoming a great historian, or anything else
above mediocrity.   But in Parkman, as in all men
of genius, the dominant motive power was something within him, something which science has not
data enough to explain.   The divine spark of
genius is something which we know only through
the acts which it excites.   In Parkman the strong
literary instinct showed itself at Chauncy Hall
School, where we find him, at fourteen years of
age, eagerly and busily engaged in the study and

practice of English composition. It was natural that tales of heroes should be especially charming at that time of life, and among Parkman's efforts were paraphrasing parts of the Æneid, and turning into rhymed verse the scene of the tournament in "Ivanhoe." From the artificial stupidity which is too often superinduced in boys by their early schooling he was saved by native genius and breezy woodland life, and his progress was rapid. In 1840, having nearly completed his seventeenth year, he entered Harvard College. His reputation there for scholarship was good, but he was much more absorbed in his own pursuits than in the regular college studies. In the summer vacation of 1841 he made a rough journey of exploration in the woods of northern New Hampshire, accompanied by one classmate and a native guide, and there he had a taste of adventure slightly spiced with hardship.

How much importance this ramble may have had one cannot say, but he tells us that "before the end of the Sophomore year my various schemes had crystallized into a plan of writing the story of what was then known as the 'Old French War,' — that is, the war that ended in the conquest of Canada; for here, as it seemed to me, the forest drama was more stirring, and the forest stage more thronged with appropriate actors, than in any other passage of our history. It was not until some years later that I enlarged the plan to include the whole course of the American conflict between France and England, or, in other words, the history of the American forest ; for this was the light in which I regarded it. My theme fascinated me, and I was haunted with wilderness images day and night." The way in which true genius works could not be more happily described.

When the great scheme first took shape in Mr. Parkman's mind, he reckoned that it would take about twenty years to complete the task. How he entered upon it may best be told in his own words : —

" The time allowed was ample ; but here he fell into a fatal error, entering on this long pilgrimage with all the vehemence of one starting on a mile heat. His reliance, however, was less on books than on such personal experience as should in some sense identify him with his theme. His natural inclinations urged him in the same direction, for his thoughts were always in the forest, whose features, not unmixed with softer images, possessed his waking and sleeping dreams, filling him with vague cravings impossible to satisfy. As fond of hardships as he was vain of enduring them, cherishing a sovereign scorn for every physical weakness or defect, deceived moreover by a rapid development of frame and sinews which flattered him with the belief that discipline sufficiently unsparing would harden him into an athlete, he slighted the precautions of a more reasonable woodcraft, tired old foresters with long marches, stopped neither for heat nor rain, and slept on the earth without a blanket." In other words, " a highly irritable organism spurred the writer to excess in a course which, with one of different temperament, would have produced a free and hardy development of such faculties and forces as he possessed." Along with the irritable organism perhaps a heritage of fierce ancestral Puritanism may have prompted him to the stoical discipline which sought to ignore the just claims of the physical body. He tells us of his undoubting faith that " to tame the Devil, it is best to take him by the horns ; " but more mature experiences made him feel less sure " of the advantages of this method of dealing with that subtle personage."

Under these conditions, perhaps the college vacations which he spent in the woods of Canada and New England may have done more to exhaust than to recruit his strength. In his Junior year, some physical injury, the nature of which does not seem to be known, caused it to be thought necessary to send him to Europe for his health. He went first to Gibraltar in a sailing ship, and a passage from his diary may serve to throw light upon the voyage and the man : " It was a noble sight when at intervals the sun broke out over the savage waves, changing their blackness to a rich blue almost as dark ; while the foam that flew over it seemed like whirling snow wreaths on the mountain. . . . As soon as it was daybreak I went on deck. Two or three sails were set. The vessel was scouring along, leaning over so that her lee gunwale scooped up the water; the water in a foam, and clouds of spray flying over us, frequently as high as the main yard. The spray was driven with such force that it pricked the cheek like needles. I stayed on deck two or three hours, when, being thoroughly salted, I went down, changed my clothes, and read ' Don Quixote ' till Mr. Snow appeared at the door with ' You are the man that wants to see a gale, are ye ? Now is your chance ; only just come up on deck.' Accordingly I went. The wind was yelling and howling in the rigging in a fashion that reminded me of a storm in a Canadian forest. . . . The sailors clung, half drowned, to whatever they could lay hold of, for the vessel was at times half inverted, and tons of water washed from side to side of her deck."

Mr. Parkman's route was from Gibraltar by way of Malta, to Sicily, where he travelled over the whole island, and thence to Naples, where he fell in with the great preacher Theodore Parker. Together they climbed Vesuvius and peered into its crater, and afterwards in and about Rome they renewed their comradeship. Here Mr. Parkman wished to spend a few weeks in a monastery, in order to study with his own eyes the priests and their way of life. More than once he met with a prompt and uncompromising refusal, but at length the coveted privilege was granted him; and, curiously enough, it was by the strictest of all the monastic orders, the Passionists, brethren addicted to wearing hair shirts and scourging themselves without mercy. When these worthy monks learned that their visitor was not merely a Protestant, but a Unitarian, their horror was intense; but they were ready for the occasion, poor souls! and tried their best to convert him, thereby doubtless enhancing their value in the historian's eyes as living and breathing historic material. This visit was surely of inestimable service to the pen which was to be so largely occupied with the Jesuits and Franciscans of the New World.

Mr. Parkman did not leave Rome until he had seen temples, churches, and catacombs, and had been presented to the Pope. He stopped at Florence, Bologna, Modena, Parma, and Milan, and admired the Lake of Como, to which, however, he preferred the savage wildness of Lake George. He saw something of Switzerland, went to Paris and London, and did a bit of sight-seeing in Edinburgh and its neighbourhood. From Liverpool he sailed for America; and in spite of the time consumed in this trip we find him taking his degree at Cambridge, along with his class, in 1844. Probably his name stood high in the rank list, for he was at once elected a member of the Phi Beta Kappa Society. After this he entered the Law School, but stayed not long, for his life's work was already claiming him. In his brief vacation journeys he had seen tiny remnants of wilderness here and there in Canada or in lonely corners of New England; now he wished to see the wilderness itself in all its gloom and vastness, and to meet face to face with the dusky warriors of the Stone Age. At this end of the nineteenth century, as already observed, such a thing can no longer be done. Nowhere now, within the United States, does the primitive wilderness exist, save here and there in shreds and patches. In the middle of the century it covered the western half of the continent, and could be reached by a journey of sixteen or seventeen hundred miles, from Boston to the plains of Nebraska. Parkman had become an adept in woodcraft and a dead shot with the rifle, and could do such things with horses, tame or wild, as civilized people never see done except in a circus. There was little doubt as to his ability to win the respect of Indians by outshining them in such deeds as they could appreciate. Early in 1846 he started for the wilderness with Mr. Quincy Shaw. A passage from the preface to the fourth edition of "The Oregon Trail," published in 1872, will here be of interest : —

"I remember, as we rode by the foot of Pike's Peak, when for a fortnight we met no face of man, my companion remarked, in a tone anything but complacent, that a time would come when those plains would be a grazing country, the buffalo give place to tame cattle, houses be scattered along the watercourses, and wolves, bears, and Indians be numbered among the things that were. We condoled with each other on so melancholy a prospect, but with little thought what the future had in store. We knew that there was more or less gold in the seams of those untrodden mountains ; but we did not foresee that it would build cities in the West, and plant hotels and gambling houses among the haunts of the grizzly bear. We knew that a few

fanatical outcasts were groping their way across
the plains to seek an asylum from Gentile persecution; but we did not imagine that the polygamous
hordes of Mormons would rear a swarming Jeru-
salem in the bosom of solitude itself. We knew
that more and more, year after year, the trains of
emigrant wagons would creep in slow procession
towards barbarous Oregon or wild and distant
California; but we did not dream how Commerce
and Gold would breed nations along the Pacific,
the disenchanting screech of the locomotive break
the spell of weird, mysterious mountains, woman's
rights invade the fastnesses of the Arapahoes, and
despairing savagery, assailed in front and rear,
veil its scalp locks and feathers before triumphant
commonplace. We were no prophets to foresee
all this; and had we foreseen it, perhaps some perverse regret might have tempered the ardour of
our rejoicing.

"The wild tribe that defiled with me down the
gorges of the Black Hills, with its paint and war
plumes, fluttering trophies and savage embroidery,
bows, arrows, lances, and shields, will never be seen
again. Those who formed it have found bloody
graves, or a ghastlier burial in the maws of wolves.
The Indian of to-day, armed with a revolver and
crowned with an old hat, cased possibly in trou-

sers or muffled in a tawdry shirt, is an Indian still, but an Indian shorn of the picturesqueness which was his most conspicuous merit. The mountain trapper is no more, and the grim romance of his wild, hard life is a memory of the past."

This first of Parkman's books, "The Oregon Trail," was published in 1847, as a series of articles in the "Knickerbocker Magazine." Its pages reveal such supreme courage, such physical hardiness, such rapturous enjoyment of life, that one finds it hard to realize that even in setting out upon this bold expedition the writer was something of an invalid. A weakness of sight — whether caused by some direct injury, or a result of widespread nervous disturbance, is not quite clear — had already become serious and somewhat alarming. On arriving at the Indian camp, near the Medicine Bow range of the Rocky Mountains, he was suffering from a complication of disorders. " I was so reduced by illness," he says, " that I could seldom walk without reeling like a drunken man; and when I rose from my seat upon the ground the landscape suddenly grew dim before my eyes, the trees and lodges seemed to sway to and fro, and the prairie to rise and fall like the swells of the ocean. Such a state of things is not enviable anywhere. In a country where a man's

life may at any moment depend on the strength
of his arm, or it may be on the activity of his legs,
it is more particularly inconvenient. Nor is sleeping on damp ground, with an occasional drenching
from a shower, very beneficial in such cases. I
sometimes suffered the extremity of exhaustion,
and was in a tolerably fair way of atoning for my
love of the prairie by resting there forever. I
tried repose and a very sparing diet. For a long
time, with exemplary patience, I lounged about
the camp, or at the utmost staggered over to the
Indian village, and walked faint and dizzy among
the lodges. It would not do, and I bethought me
of starvation. During five days I sustained life on
one small biscuit a day. At the end of that time
I was weaker than before, but the disorder seemed
shaken in its stronghold, and very gradually I began to resume a less rigid diet." It did not seem
prudent to Parkman to let the signs of physical
ailment become conspicuous, " since in that case
a horse, a rifle, a pair of pistols, and a red shirt
might have offered temptations too strong for
aboriginal virtue." Therefore, in order that his
prestige with the red men might not suffer diminution, he would "hunt buffalo on horseback over
a broken country, when without the tonic of the
chase he could scarcely sit upright in the saddle."

The maintenance of prestige was certainly desirable. The Ogillalah band of Sioux, among whom he found himself, were barbarians of a low type. "Neither their manners nor their ideas were in the slightest degree modified by contact with civilization. They knew nothing of the power and real character of the white men, and their children would scream in terror when they saw me. Their religion, superstitions, and prejudices were the same handed down to them from immemorial time. They fought with the weapons that their fathers fought with, and wore the same garments of skins. They were living representatives of the Stone Age ; for, though their lances and arrows were tipped with iron procured from the traders, they still used the rude stone mallet of the primeval world." These savages welcomed Parkman and one of his white guides with cordial hospitality, and they were entertained by the chieftain Big Crow, whose lodge in the evening presented a picturesque spectacle. " A score or more of Indians were seated around it in a circle, their dark, naked forms just visible by the dull light of the smouldering fire in the middle. The pipe glowed brightly in the gloom as it passed from hand to hand. Then a squaw would drop a piece of buffalo fat on the dull embers. Instantly a bright flame would leap up,

darting its light to the very apex of the tall coni-
cal structure, where the tops of the slender poles
that supported the covering of hide were gathered
together. It gilded the features of the Indians,
as with animated gestures they sat around it, telling their endless stories of war and hunting, and
displayed rude garments of skins that hung around
the lodge ; the bow, quiver, and lance suspended
over the resting place of the chief, and the rifles
and powderhorns of the two white guests. For
a moment all would be bright as day; then the
flames would die out; fitful flashes from the embers
would illumine the lodge, and then leave it in darkness. Then the light would wholly fade, and the
lodge and all within it be involved again in ob-
scurity." From stories of war and the chase the
conversation was now and then diverted to philosophic themes. When Parkman asked what makes
the thunder, various opinions were expressed ; but
one old wrinkled fellow, named Red Water, assev-
erated that he had always known what it was.
" It was a great black bird; and once he had seen
it in a dream swooping down from the Black Hills,
with its loud roaring wings ; and when it flapped
them over a lake, they struck lightning from the
water." Another old man said that the wicked
thunder had killed his brother last summer, but

doggedly refused to give any particulars. It was
afterwards learned that this brother was a member of a thunder-fighting fraternity of priests or
medicine men.   On the approach of a storm they
would " take their bows and arrows, their magic
drum, and a sort of whistle made out of the wing
bone of the war eagle, and, thus equipped, run out
and fire at the rising cloud, whooping, yelling,
whistling, and beating their drum, to frighten it
down again.   One afternoon a heavy black cloud
was coming up, and they repaired to the top of a
hill, where they brought all their magic artillery
into play against it.   But the undaunted thunder,
refusing to be terrified, darted out a bright flash,
which struck [the aforesaid brother] dead as he
was in the very act of shaking his long iron-pointed
lance against it.   The rest scattered, and ran yelling in an ecstasy of superstitious terror back to
their lodges."

One should read Mr. Parkman's detailed narrative of the strange life of these people, and the
manner of his taking part in it : how he called the
villagers together and regaled them sumptuously
with boiled dog, and made them a skilful speech,
in which he quite satisfied them as to his reasons
for coming to dwell among them ; how a warm
friendship grew up between himself and the ven-

erable Red Water, who was the custodian of an
immense fund of folk lore, but was apt to be super-
stitiously afraid of imparting any of it to strangers;
how war parties were projected and abandoned;
how buffalo and antelope were hunted, and how
life was carried on in the dull intervals between
such occupations. If one were to keep on quoting
what is of especial interest in the book, one would
have to quote the whole of it. But one characteristic portrait contains so much insight into Indian
life that I cannot forbear giving it. It is the
sketch of a young fellow called the Hail-Storm,
as Parkman found him one evening on his return
from the chase: " his light graceful figure reclining
on the ground in an easy attitude, while . . . near
him lay the fresh skin of a female elk which he
had just killed among the mountains, only a mile
or two from camp. No doubt the boy's heart was
elated with triumph, but he betrayed no sign of
it. He even seemed totally unconscious of our
approach, and his handsome face had all the tran-
quillity of Indian self-control, — a self-control which
prevents the exhibition of emotion without restrain-
ing the emotion itself. It was about two months
since I had known the Hail-Storm, and within
that time his character had remarkably developed.
When I first saw him, he was just emerging from

the habits and feelings of the boy into the ambition
of the hunter and warrior.　He had lately killed
his first deer, and this had excited his aspirations
for distinction.　Since that time he had been continually in search for game, and no young hunter
in the village had been so active or so fortunate
as he.　All this success had produced a marked
change in his character.　As I first remembered
him, he always shunned the society of the young
squaws, and was extremely bashful and sheepish
in their presence ; but now, in the confidence of
his new reputation, he began to assume the airs
and arts of a man of gallantry.　He wore his red
blanket dashingly over his left shoulder, painted
his cheeks every day with vermilion, and hung pen-
dants of shells in his ears.　If I observed aright,
he met with very good success in his new pursuits ;
still the Hail-Storm had much to accomplish before
he attained the full standing of a warrior.　Gal-
lantly as he began to bear himself before the women and girls, he was still timid and abashed in
the presence of the chiefs and old men ; for he had
never yet killed a man, or stricken the dead body
of an enemy in battle.　I have no doubt that the
handsome smooth-faced boy burned with desire to
flesh his maiden scalping knife, and I would not
have encamped alone with him without watching

his movements with a suspicious eye." Mr. Parkman once told me that it was rare for a young brave to obtain full favour with the women without having at least one scalp to show; and this fact was one of the secret sources of danger which the ordinary white visitor would never think of. Peril is also liable to lurk in allowing one's self to be placed in a ludicrous light among these people; accordingly, whenever such occasions arose, Parkman knew enough to "maintain a rigid, inflexible countenance, and [thus] wholly escaped their sallies." He understood that his rifle and pistols were the only friends on whom he could invariably rely when alone among Indians. His own observation taught him "the extreme folly of confidence, and the utter impossibility of foreseeing to what sudden acts the strange, unbridled impulses of an Indian may urge him. When among this people, danger is never so near as when you are unprepared for it, never so remote as when you are armed and on the alert to meet it at any moment. Nothing offers so strong a temptation to their ferocious instincts as the appearance of timidity, weakness, or security."

The immense importance of this sojourn in the wilderness, in its relation to Parkman's life work, is obvious. Knowledge, intrepidity, and tact carried

him through it unscathed, and good luck kept him
clear of encounters with hostile Indians, in which
these qualities might not have sufficed to avert destruction. It was rare good fortune that kept his
party from meeting with an enemy during five
months of travel through a dangerous region.
Scarcely three weeks after he had reached the con-
fines of civilization, the Pawnees and Comanches
began a systematized series of hostilities, and "at-
tacked . . . every party, large or small, that passed
during the next six months."

During this adventurous experience, says Parkman, "my business was observation, and I was
willing to pay dearly for the opportunity of exercising it." A heavy price was exacted of him, not
by red men, but by that "subtle personage" whom
he had tried to take by the horns, and who seems
to have resented such presumption. Toward the
end of the journey Parkman found himself ill
in much the same way as at the beginning, and
craved medical advice. It was in mid-September,
on a broad meadow in the wild valley of the Arkansas, where his party had fallen in with a huge
Santa Fé caravan of white-topped wagons, with
great droves of mules and horses; and we may let
Parkman tell the story in his own words, in the
last of our extracts from his fascinating book. One

of the guides had told him that in this caravan was
a physician from St. Louis, by the name of Dobbs,
of the very highest standing in his profession.
"Without at all believing him, I resolved to consult this eminent practitioner. Walking over to
the camp, I found him lying sound asleep under
one of the wagons. He offered in his own person
but indifferent evidence of his skill; for it was five
months since I had seen so cadaverous a face.
His hat had fallen off, and his yellow hair was all
in disorder ; one of his arms supplied the place of
a pillow; his trousers were wrinkled halfway up
to his knees, and he was covered with little bits
of grass and straw upon which he had rolled in his
uneasy slumber. A Mexican stood near, and I
made him a sign to touch the doctor. Up sprang
the learned Dobbs, and sitting upright rubbed his
eyes and looked about him in bewilderment. I
regretted the necessity of disturbing him, and said
I had come to ask professional advice.

" ' Your system, sir, is in a disordered state,'
said he solemnly, after a short examination. I
inquired what might be the particular species of
disorder. ' Evidently a morbid action of the liver,'
replied the medical man. ' I will give you a pre-
scription.'

" Repairing to the back of one of the covered

wagons, he scrambled in; for a moment I could see nothing of him but his boots. At length he produced a box which he had extracted from some dark recess within, and opening it presented me with a folded paper. 'What is it?' said I. 'Calomel,' said the doctor.

"Under the circumstances I would have taken almost anything. There was not enough to do me much harm, and it might possibly do good; so at camp that night I took the poison instead of supper."

After the return from the wilderness Parkman found his physical condition rather worse than better. The trouble with the eyes continued, and we begin to find mention of a lameness which was sometimes serious enough to confine him to the house, and which evidently lasted a long time; but from this he seems to have recovered. My personal acquaintance with him began in 1872, and I never noticed any symptoms of lameness, though I remember taking several pleasant walks with him. Perhaps the source of 'the lameness may be indicated in the following account of his condition in 1848, cited from the fragment of autobiography in which he uses the third person: "To the maladies of the prairie succeeded a suite of exhausting disorders, so reducing him that circulation of the extremities ceased, the light of the sun became insupportable, and a wild whirl possessed his brain, joined to a universal turmoil of the nervous system which put his philosophy to the sharpest test it had hitherto known. All collapsed, in short, but the tenacious strength of muscles hardened by long activity." In 1851, whether due or not to disordered circulation, there came an effusion of water on the left knee, which for the next two years prevented walking.

It was between 1848 and 1851 that Parkman was engaged in writing "The Conspiracy of Pontiac." He felt that no regimen could be worse for him than idleness, and that no tonic could be more bracing than work in pursuance of the lofty purpose which had now attained maturity in his mind. He had to contend with a "triple-headed monster: " first, the weakness of the eyes, which had come to be such that he could not keep them open to the light while writing his own name ; secondly, the incapacity for sustained attention ; and thirdly, the indisposition to putting forth mental effort. Evidently, the true name of this triple-headed monster was nervous exhaustion ; there was too much soul for the body to which it was yoked.

"To be made with impunity, the attempt must be made with the most watchful caution. He

caused a wooden frame to be constructed of the
size and shape of a sheet of letter paper. Stout
wires were fixed horizontally across it, half an inch
apart, and a movable back of thick pasteboard fitted
behind them. The paper for writing was placed
between the pasteboard and the wires, guided by
which, and using a black lead crayon, he could
write not illegibly with closed eyes. He was at the
time absent from home, on Staten Island, where,
and in the neighbouring city of New York, he had
friends who willingly offered their aid. It is need-
less to say to which half of humanity nearly all these
kind assistants belonged. He chose for a beginning that part of the work which offered fewest difficulties and with the subject of which he was most
familiar; namely, the Siege of Detroit. The books
and documents, already partially arranged, were procured from Boston, and read to him at such times as
he could listen to them; the length of each reading
never without injury much exceeding half an hour,
and periods of several days frequently occurring
during which he could not listen at all. Notes were
made by him with closed eyes, and afterwards de-
ciphered and read to him till he had mastered them.
For the first half-year the rate of composition aver-
aged about six lines a day. The portion of the book
thus composed was afterwards partially rewritten.

" His health improved under the process, and
the remainder of the volume — in other words,
nearly the whole of it — was composed in Boston,
while pacing in the twilight of a large garret, the
only exercise which the sensitive condition of his
sight permitted him in an unclouded day while the
sun was above the horizon. It was afterwards
written down from dictation by relatives under the
same roof, to whom he was also indebted for the
preparatory readings. His progress was much less
tedious than at the outset, and the history was
complete in about two years and a half."

The book composed under such formidable difficulties was published in 1851. It did not at
once meet with the reception which it deserved.
The reading public did not expect to find enter-
tainment in American history. In the New England of those days the general reader had heard a
good deal about the Pilgrim Fathers and Salem
Witchcraft, and remembered hazily the stories of
Hannah Dustin and of Putnam and the wolf, but
could not be counted on for much else before the
Revolution. I remember once hearing it said that
the story of the " Old French War " was something of no more interest or value for Americans
of to-day than the cuneiform records of an insur-
rection in ancient Nineveh ; and so slow are peo-

ple in gaining a correct historical perspective that
within the last ten years the mighty world struggle in which Pitt and Frederick were allied is
treated in a book entitled " Minor Wars of the
United States " !   In 1851 the soil was not yet
ready for the seed sown by Parkman, and he did
not quickly or suddenly become popular.   But
after the publication of the " Pioneers of France "
in 1865 his fame grew rapidly.   In those days I
took especial pleasure in praising his books, from
the feeling that they were not so generally known
as they ought to be, particularly in England,
where he has since come to be recognized as foremost among American writers of history.   In 1879
I had been giving a course of lectures at University College, London, on " America's Place in History," and shortly afterwards repeated this course
at the little Hawthorne Hall, on Park Street, in
Boston.   One evening, having occasion to allude
briefly to Pontiac and his conspiracy, I said, among
other things, that it was memorable as " the theme
of one of the most brilliant and fascinating books
that have ever been written by any historian since
the days of Herodotus."   The words were scarcely
out of my mouth when I happened to catch sight
of Mr. Parkman in my audience.   I had not observed him before, though he was seated quite

near me. I shall never forget the sudden start which he gave, and the heightened colour of his noble face, with its curious look of surprise and pleasure, — an expression as honest and simple as one might witness in a rather shy schoolboy suddenly singled out for praise. I was so glad that I had said what I did without thinking of his hearing me.

In May, 1850, while at work upon this great book, Mr. Parkman married Catherine, daughter of Jacob Bigelow, an eminent physician of Boston. Of this marriage there were three children, — a son, who died while an infant, and two daughters, who still survive. Mrs. Parkman died in 1858, and her husband never married again.

During these years, when his complicated ailments for a time made historical work impossible even to this man of Titanic will, he assuaged his cravings for spiritual creation by writing a novel, " Vassall Morton." Of his books it is the only one that I have never seen, and I can speak of it only from hearsay. It is said to be not without signal merits, but it did not find a great many readers, and its author seems not to have cared much for it. The main current of his interest in life was too strong to allow of much diversion into side channels.

" Meanwhile," to cite his own words, " the Faculty of Medicine were not idle, displaying that exuberance of resource for which that remarkable profession is justly famed. The wisest, indeed, did nothing, commending his patient to time and faith; but the activity of his brethren made full amends for this masterly inaction. One was for tonics, another for a diet of milk; one counselled galvanism, another hydropathy; one scarred him behind the neck with nitric acid, another drew red-hot irons along his spine with a view of enlivening that organ. Opinion was divergent as practice. One assured him of recovery in six years; another thought that he would never recover. Another, with grave circumlocution, lest the patient should take fright, informed him that he was the victim of an organic disease of the brain which must needs dispatch him to another world within a twelvemonth; and he stood amazed at the smile of an auditor who neither cared for the announcement nor believed it. Another, an eminent physiologist of Paris, after an acquaintance of three months, one day told him that from the nature of the disorder he had at first supposed that it must, in accordance with precedent, be attended with insanity, and had ever since been studying him to discover under what form the supposed aberration declared itself; adding, with a somewhat humorous look, that his researches had not been rewarded with the smallest success."

Soon after his marriage Mr. Parkman became possessor of a small estate of three acres or so in Jamaica Plain, on the steep shore of the beautiful pond. It was a charming place, thoroughly English in its homelike simplicity and refined comfort. The house stood near the entrance, and on not far from the same level as the roadway; but from the side and rear the ground fell off rapidly, so that it was quite a sharp descent to the pretty little wharf or dock, where one might sit and gaze on the placid, dreamy water. It is with that lovely home that Parkman is chiefly associated in my mind. Twenty years ago, while I was acting as librarian at Harvard University, he was a member of the corporation, and I had frequent occasion to consult with him on matters of business. At such times I would drive over from Cambridge or take a street car to Jamaica Plain, sure of a cordial greeting and a pleasant chat, in which business always received its full measure of justice, and was then thrust aside for more inspiring themes. The memory of one day in particular will go with me through life, — an enchanted day in the season of apple blossoms, when I went in the morning for a brief errand, taking with me one of my little sons. The brief errand ended in spending the whole day and staying until late in the evening, while the world of thought was ransacked and some of its weightiest questions provisionally settled ! Nor was either greenhouse or garden or pond neglected. At such times there was nothing in Parkman's looks or manner to suggest the invalid. He and I were members of a small club of a dozen or more congenial spirits who now for nearly thirty years have met once a month to dine together. When he came to the dinner he was always one of the most charming companions at the table; but ill health often prevented his coming, and in the latter years of his life he never came. I knew nothing of the serious nature of his troubles; and when I heard the cause of his absence alleged, I used to suppose that it was merely some need for taking care of digestion or avoiding late hours that kept him at home. What most impressed one, in talking with him, was the combination of power and alertness with extreme gentleness. Nervous irritability was the last thing of which I should have suspected him. He never made the slightest allusion to his ill health; he would probably have deemed it inconsistent with good breeding to intrude upon his

friends with such topics; and his appearance was
always most cheerful. His friend (our common
friend), the late Octavius Frothingham, says of
him : " Again and again he had to restrain the
impulse to say vehement things, or to do violent
deeds without the least provocation ; but he maintained so absolutely his moral self-control that
none but the closest observer would notice any
deviation from the most perfect calm and serenity." I can testify that until after Mr. Parkman's death I had never dreamed of the existence
of any such deviation.

Garden and greenhouse formed a very important part of the home by Jamaica Pond. Mr.
Parkman's love for Nature was in no way more
conspicuously shown than in his diligence and
skill in cultivating flowers. It is often observed
that plants will grow for some persons, but not
for others ; one man's conservatory will be heavy
with verdure, gorgeous in its colours, and redolent
of sweet odours, while his neighbour's can show
nothing but a forlorn assemblage of pots and sticks.
The difference is due to the loving care which
learns and humours the idiosyncrasies of each individual thing that grows, the keen observation
of the naturalist supplemented by the watchful
solicitude of the nurse. Among the indications

of rare love and knowledge of Nature is marked success in inducing her to bring forth her most exquisite creations, the flowers. As an expert in horticulture Parkman achieved celebrity. His garden and greenhouse had extraordinary things to show. As he pointed out to me on my first visit to them, he followed Darwinian methods and originated new varieties of plants. The *Lilium Parkmani* has long been famous among florists. He was also eminent in the culture of roses, and author of a work entitled "The Book of Roses," which was published in 1866. He was President of the Horticultural Society, and at one time Professor of Horticulture in Harvard University. There can be no doubt as to the beneficial effects of these pursuits. It is wholesome to be out of doors with spade and trowel and sprinkler ; there is something tonic in the aroma of fresh damp loam ; and nothing is more restful to the soul than daily sympathetic intercourse with flowering plants. It was surely here that Parkman found his best medicine.

When he entered, in 1851, upon his great work on " France and England in the New World," he had before him the task " of tracing out, collecting, indexing, arranging, and digesting a great mass of incongruous material scattered on both sides of the

Atlantic." A considerable portion of this material
was in manuscript, and involved much tedious exploration and the employment of trained copyists.
It was necessary to study carefully the catalogues
of many European libraries, and to open correspondence with such scholars and public officials in both
hemispheres as might be able to point to the whereabouts of fresh sources of information. Work of
this sort, as one bit of clue leads to another, is capable of arousing the emotion of pursuit to a very
high degree; and I believe the effect of it upon
Parkman's health must have been good, in spite
of, or rather because of, its difficulties. The
chase was carried on until his manuscript treasures had been brought to an extraordinary degree of completeness. These made his library
quite remarkable. In printed books it was far less
rich. He had not the tastes of a bibliophile, and
did not feel it necessary, as Freeman did, to own
all the books he used. His library of printed
books, which at his death went to Harvard University, was a very small one for a scholar, — about
twenty-five hundred volumes, including more or
less of Greek and Latin literature and theology
inherited from his father. His manuscripts, as I
have already mentioned, went to the library of
the Massachusetts Historical Society.

When the manuscripts had come into his hands,
an arduous labour was begun.   All had to be read
to him and taken in slowly, bit by bit.   The incapacity to keep steadily at work made it impossible
to employ regular assistants profitably ; and for
readers he either depended upon members of his
own family or called in pupils from the public
schools.   Once he speaks of having had a well-
trained young man, who was an excellent linguist ;
on another occasion it was a schoolgirl " ignorant
of any tongue but her own," and " the effect, though
highly amusing to bystanders, was far from being
so to the person endeavouring to follow the meaning
of this singular jargon."   The larger part of the
documents used in preparing the earlier volumes
were in seventeenth-century French, which, though
far from being Old French, is enough unlike the
nineteenth-century speech to have troubled Parkman's readers, and thus to have worried his ears.

As Frothingham describes his method, when
the manuscripts were slowly read to him, " first
the chief points were considered, then the details
of the story were gone over carefully and minutely.
As the reading went on he made notes, first of essential matters, then of non-essential.   After this
he welded everything together, made the narrative
completely his own, infused into it his own fire,

quickened it by his own imagination, and made it, as it were, a living experience, so that his books read like personal reminiscences. It was certainly a slow and painful process, but the result more than justified the labour."

In the fragment of autobiography already quoted, which Mr. Parkman left with Dr. Ellis in 1868, but which was apparently written in 1865, he says: " One year, four years, and numerous short intervals lasting from a day to a month represent the literary interruptions since the work in hand was begun. Under the most favourable conditions it was a slow and doubtful navigation, beset with reefs and breakers, demanding a constant lookout and a constant throwing of the lead. Of late years, however, the condition of the sight has so far improved as to permit reading, not exceeding on the average five minutes at one time. This modicum of power, though apparently trifling, proves of the greatest service, since by a cautious management its application may be extended. By reading for one minute, and then resting for an equal time, this alternate process may generally be continued for about half an hour. Then after a sufficient interval it may be repeated, often three or four times in the course of the day. By this means nearly the whole of the volume now offered [" Pioneers "] has been

composed. . . . How far, by a process combining
the slowness of the tortoise with the uncertainty
of the hare, an undertaking of close and extended
research can be advanced, is a question to solve
which there is no aid from precedent, since it does
not appear that an attempt under similar circumstances has hitherto been made. The writer looks,
however, for a fair degree of success."

After 1865 the progress was certainly much
more rapid than before. The next fourteen years
witnessed the publication of " The Jesuits," " La
Salle," " The Old Régime," and " Frontenac,"
and saw " Montcalm and Wolfe " well under way ;
while the " Half-Century of Conflict," interven-
ing between " Frontenac " and " Montcalm and
Wolfe," was reserved until the last-mentioned work
should be done, for the same reason that led Herbert Spencer to postpone the completing of his
" Sociology" until he should have finished his
" Principles of Ethics." In view of life's vicissitudes, it was prudent to make sure of the crown-
ing work, at all events, leaving some connecting
links to be inserted afterwards. As one obstacle
after another was surmounted, as one grand division of the work after another became an accomplished fact, the effect upon Parkman's condition must have been bracing, and he seems to

have acquired fresh impetus as he approached the goal.

For desultory work in the shape of magazine articles he had little leisure ; but two essays of his, on " The Failure of Universal Suffrage " and on " The Reasons against Woman Suffrage," are very thoughtful, and worthy of serious consideration. In questions of political philosophy, his conclusions, which were reached from a very wide and impartial survey of essential facts, always seemed to me of the highest value.

When I look back upon Parkman's noble life, I think of Mendelssohn's chorus, " He that shall endure to the end," with its chaste and severely beautiful melody, and the calm, invincible faith which it expresses. After all the harrowing years of doubt and distress, the victory was such in its magnitude as has been granted to but few mortals to win. He lived to see his life's work done ; the thought of his eighteenth year was realized in his sixty-ninth ; and its greatness had come to be admitted throughout the civilized world. In September, 1893, his seventieth year was completed, and his autumn in the lovely home at Jamaica Plain was a pleasant one. On the first Sunday afternoon in November he rowed on the pond in his boat, but felt ill as he returned to the house, and on the next Wednesday, the 8th, he passed quietly away. Thus he departed from a world which will evermore be the richer and better for having once had him as its denizen. The memory of a life so strong and beautiful is a precious possession for us all.

As for the book on which he laboured with such marvellous heroism, a word may be said in conclusion. Great in his natural powers and great in the use he made of them, Parkman was no less great in his occasion and in his theme. Of all American historians he is the most deeply and peculiarly American, yet he is at the same time the broadest and most cosmopolitan. The book which depicts at once the social life of the Stone Age, and the victory of the English political ideal over the ideal which France inherited from imperial Rome, is a book for all mankind and for all time. The more adequately men's historic perspective gets adjusted, the greater will it seem. Strong in its individuality, and like to nothing else, it clearly belongs, I think, among the world's few masterpieces of the highest rank, along with the works of Herodotus, Thucydides, and Gibbon.

February, 1897.

IX

EDWARD AUGUSTUS FREEMAN

THE sudden death of Professor Freeman, last March [1892], was a great calamity to the world of letters. Although his achievements in the field of historical writing had been so varied and voluminous, yet some of his most important themes — some of those which had been slowly ripening and most richly developed in his mind — were still awaiting literary treatment at his hands, and at the time of his death he had just finished the third volume of a colossal work which was still in its earlier stages. His end was premature, and it is with a keen sense of bereavement that we take this occasion to pay a brief word of tribute to so dear and honoured a teacher.

Edward Augustus Freeman, son of John Freeman of Redmore Hall, in Worcestershire, was born at Harborne, Staffordshire, August 2, 1823. His life was always purely that of a scholar and teacher, and a chronicle of its events would consist chiefly of the record of books published and offices held

at the University of Oxford. He was graduated
at Trinity College in 1845, and remained there as
a Fellow until 1847. In 1857, 1863, and 1873
he served as Examiner in Modern History. In
1880 he was chosen honorary Fellow of Trinity,
and in 1884 Fellow of Oriel. In the latter year
he was appointed Regius Professor of Modern History, succeeding Bishop Stubbs in that position.
It is not necessary to enumerate the honorary
degrees which he received from Oxford and Cambridge, and from universities in various European
countries. At the time of his death he was a member of learned societies in nearly all parts of the
world. For many years he had been a Knight
Commander of the Greek Order of the Saviour.
He had also received honours of knighthood from
Servia and Montenegro. In 1868 he was a candidate for Parliament, but failed of election; and
that seems to have been his sole venture in the
world of politics. His travels upon the continent
of Europe were many and extensive. When at
home he lived in rural seclusion, — " far from the
madding crowd," — upon his estate at Somerleaze,
near Wells and its noble cathedral; only in these
latter years he made a home for himself, during
the Oxford terms, at St. Giles in that city.

From the very beginning Freeman's histori-

cal studies were characterized on the one hand by
philosophical breadth of view, and on the other
hand by extreme accuracy of statement, and such
loving minuteness of detail as is apt to mark the
local antiquary whose life has been spent in studying only one thing. It was to the combination of
these two characteristics that the preëminent greatness of his historical work was due. We see the
combination already prefigured, and to some extent
realized, in his first book, " A History of Architecture," published in 1849, although this can hardly
be called such a work of original research as the
books of his maturer years. Two years afterward
appeared the learned " Essay on the Origin and Development of Window Tracery in England," a work
which I do not feel able to criticise, but which I
am sure is very charming to read. I believe that
this book was followed by at least three others in
the same department, " Architectural Antiquities of
Gower," " The Antiquities of St. David's," and
" The Architecture of Llandaff Cathedral," but I
have never seen them. In the preface to the essay on
window tracery Mr. Freeman alludes to Rev. G. W.
Cox as his " friend and coadjutor in many undertakings," and I have heard of a volume of poems
" by G. W. C. and E. A. F." published in those days,
but I know no more about it. It is to be hoped that

these early works, which have become very scarce, will before long be collected and reprinted.

When, after these publications on architecture, Freeman began publishing books and articles on ancient Greece and on the Saracens, I presume there were many of his readers who thoughtlessly assumed that he had changed his vocation; he must more than once have had to answer the stupid question why he had gone over from architecture to history. But in his mind the evolution of architecture was never separated from the course of political history; and the effect of these early studies in architecture, which were indeed never abandoned, but kept up with enthusiasm in later years, was to give increased definiteness and concreteness to his presentation of historical events. When I use such a word as "evolution" in this connection, I do not mean that Mr. Freeman was in any sense a "disciple" of the modern evolution philosophy. There is nothing to show that he ever gave any time or attention to the study of that subject, or that he had any technical knowledge even of its terminology. Whether consciously or unconsciously, however, he was an evolutionist in spirit. From the outset he was deeply impressed with the solidarity of human history, and no student of political development in our time has made more effective use of the comparative method.

From 1850 to 1863 Freeman's published writings were chiefly concerned with Mediterranean history viewed on the broadest scale in relation to all those movements of progressive humanity which have had that great inland sea for a common centre. Here came those brilliant essays on " Ancient Greece and Mediæval Italy," " Homer and the Homeric Age," " The Athenian Democracy," " Alexander the Great," " Greece during the Macedonian Period," " Mommsen's History of Rome," " The Flavian Cæsars," and others since collected in the second series of his " Historical Essays." To this period also belongs the little book on the " History of the Saracens," based upon lectures given at the Philosophical Institution in Edinburgh.

From these Mediterranean studies may be said to have grown two of Freeman's three great works, — both of them, unfortunately, left incomplete at his death, — the " History of Federal Government " and the " History of Sicily." Freeman was remarkably free from the common habit — common even among eminent historians — of concentrating his attention upon some exceptionally brilliant period or so-called " classical age," to the exclusion of other ages that went before and came after. Such a habit is fatal to all correct understanding of history, even that of the ages upon

which attention is thus unwisely concentrated.
Freeman understood that in some respects, if not
in others, the history of Greece is just as important after the battle of Chæronea as before ; and
he became especially interested in the history of
the Achaian League and other Greek attempts
at federation. Thence grew the idea of studying
the development of federal union as the highest
form of nation-building, beginning with its germs
in the leagues among Greek autonomous cities.
The enterprise was arduous, involving as it did the
determination of obscure points in the history of
many ages and countries, more particularly Greece,
Switzerland, and America. The first volume, containing the general introduction and the history of
the Greek federations, was published in 1863, a stal-
wart octavo of 721 pages. It bore upon the title-
page a motto from " The Federalist," No. XVIII., —
" Could the interior structure and regular operation
of the Achaian League be ascertained, it is probable that more light might be thrown by it on the
science of federal government than by any of the
like experiments with which we are acquainted."
This book is of priceless value, and if Freeman
had never published anything more, it would have
entitled him to a place in the foremost rank of
historians. It deals thoroughly with a very im-

portant portion of the world's history to which no
one before had even begun to do justice.   Its admirable philosophical spirit is matched by its keen
critical insight and its minute and exhaustive control of all sources of information.   Its narrative,
moreover, is full of human interest.   Yet it never
became a popular book.   It was hard to make
people believe that the Achaian League could be
interesting, and in order to realize the philosophical value of the whole story most readers would
need to have the later portions of it set before
their eyes.

But this noble work, in some respects the grandest of the author's conceptions, was never completed.   The first volume was all that ever was
published.   For this fact I have sometimes heard
Americans offer a grotesque explanation.   The volume published in 1863, in the middle of our Civil
War, bore the title " History of Federal Government, from the Foundation of the Achaian League
to the Disruption of the United States."   This title
gave offence in America.   It was too hastily taken
to indicate that the author wished well to the
Southern Confederacy, and regarded its independence as an accomplished fact.   There can be no
doubt that the title was ill chosen ; but to suppose,
as some people did, that chagrin at the success of

the Union arms prevented Freeman from going
on with his book was simply ridiculous. It was
not anything that happened in America, but something that happened in Europe, which caused
him to defer the completion of his second volume.
That volume was to deal with federal government
as exemplified in Switzerland and otherwise in Germany; and the war of 1866 between Prussia and
Austria marked the beginning of organic changes
in Germany which Freeman was anxious to watch
for a while before finishing his book.

He therefore turned aside and took up the third
of his three great works, — the only one that he
lived to complete, — the " History of the Norman
Conquest of England, its Causes and its Results."
Upon this subject he had thought and studied for
nearly twenty years, or ever since the time when
he was publishing works on architecture. As one
turns the leaves of these stout volumes, each of
seven or eight hundred pages, crowded with minute
and accurate erudition, one marvels that the author
could carry along so many researches and of such
exhaustive character at the same time. Alike in
Greek, in German, and in English history, along
with abundant generalizations, often highly original
and suggestive, we find investigations of obscure
points in which every item of evidence is weighed as

in an apothecary's scale, and in all these directions
Freeman was working at once. When it came
to publishing, volume followed volume with surprising quickness. Turning aside in 1866 from
the second volume of the " Federal Government "
when a large part of it was already written, Freeman brought out the first volume of the " Norman Conquest " in 1867, the second in 1868, the
third in 1869, the fourth in 1871, the fifth more
leisurely in 1876. The proportions of this work
are eminently characteristic of the author's historical perspective. In order to understand the
Norman Conquest, a survey of all previous English history, and especially of the struggle between
Englishmen and Danes, is essential; and the first
volume carries us in one great sweep from the land-
ing of Hengist to the accession of Edward the Con-
fessor, while the early history of Normandy also
receives due attention. We now enter the region
of proximate causes, which require more detailed
specification, and the second volume takes us
through the four-and-twenty years of Edward's
reign. His death hurries the situation to its dramatic climax, and the whole of the third volume is
devoted to the events of the single year 1066.
The completion of the Conquest down to the death
of the Conqueror is treated with less detail, and

the twenty-one years are comprised within a volume. Finally, in summing up the results of the great event, the last volume covers two centuries, and leaves us in the reign of Edward I., the king who did so much to make modern English history the glorious tale that it has been. In finishing his work upon these proportions, Freeman encountered many points in the reign of William Rufus that needed fuller treatment, and so in 1882 he published in two volumes the history of that reign as a sequel to the " Norman Conquest." Taken as a whole, the seven volumes give us such a masterly philosophic analysis and such a picturesque and vivid narrative of the history of England in the eleventh century that it must be pronounced the monumental work upon which Freeman's reputation will chiefly rest.

While these volumes were in course of publication, there was scarcely a year when its busy author, from his wealth of knowledge, did not bring out some other book. Sometimes it was what men count a slight affair, such as a textbook, — albeit the textbook is perhaps the hardest kind of book to write well; sometimes it was a brief monograph or course of lectures; sometimes a collection of earlier writings. There was an " Old English History for Children " (1869), a " Short History of the

Norman Conquest" (1880), and a " General Sketch
of European History " (1873). The " Growth of
the English Constitution " was suggestively treated
in a small volume (1872). There was a " History
of the Cathedral Church at Wells " (1870), and
there was a collection of " Historical and Architectural Sketches," chiefly from Italy (1876), followed
by " Sketches from the Subject and Neighbour
Lands of Venice " (1881). In these two last-
named volumes, illustrated chiefly from the author's
own drawings, one sees that his interest in Diocle-
tian and Theodoric was scarcely less keen than in
Alfred of Wessex or William the Norman. No
other modern traveller has done such justice to Istria
and Dalmatia. " I am not joking," he writes, " when
I say that the best guide to those parts is still the
account written by the Emperor Constantine Por-
phyrogenitus more than nine hundred years back.
But it is surely high time that there should be another." Freeman's accurate knowledge of south-
eastern Europe and its peoples, coupled with his
wide and comprehensive study of the contact between Christians and Mussulmans in all ages, led
him to take very sound and wholesome views of the
unspeakable Turk and the everlasting Eastern Question; and in 1877, when public attention was so
strongly directed toward the Balkans, he published

a lucid and graphic little volume on "The Ottoman Power in Europe." This book was a companion to the "History of the Saracens," above mentioned, and the two together make as good an introduction to Mussulman history in its relations to Europe as the general reader is likely to find.

Among the host of side works which were issued during these years, two call for especial mention. In the lectures on "Comparative Politics," given at the Royal Institution in 1873, Freeman analyzed and described the different forms assumed by Aryan institutions among Greeks, Romans, and Teutons. This book is his most distinct attempt to make his central theme the career of an institution, such as kingship or representative assemblies, rather than the career of a state or a people. In the "History of Federal Government," the two kinds of treatment, analytical and synthetical, were combined in a way that would, I think, have made that his grandest work, had it been completed. In the lectures we get an able analysis and comparison, full of fruitful suggestions, and in our author's happiest style. There is not the originality of scholarship here that we find in Sir Henry Maine, nor do we find the breadth of view that can be gained only when the barbaric non-Aryan world is taken into account. Such breadth was not to be expected twenty years ago, and before the path-breaking work of the American scholar Lewis Morgan. Freeman's outlook was confined to the Aryan domain; but he did not attempt more than he knew. His task was conceived with so clear a consciousness of his limitations, and every point was so richly illustrated, that the " Comparative Politics " remains one of his most useful and charming books.

The other work calling for especial mention is "The Historical Geography of Europe," published in 1880. Its object was " to trace out the extent of territory which the different states and nations of Europe have held at different times in the world's history ; to mark the different boundaries which the same country has had, and the different meanings in which the same name has been used." Such work is of great and fundamental importance, because men are perpetually making grotesque mistakes through ignorance or forgetfulness of the changes which have occurred upon the map ; as, for example, when somebody speaks of Lyons in the twelfth century as a French city, or supposes that Charles the Bold invaded Swiss territory. Historical writings fairly swarm with blunders based upon unconscious errors of this sort, and nowhere did Freeman do better service than in pointing

them out on every possible occasion. No writer
has so effectively warned the historical student
against that besetting sin of "bondage to the
modern map." His exposition of historical geography is a book of purest gold, and no serious
student of history can safely neglect it.

In 1881 Mr. Freeman visited the United States,
and gave lectures on "The English People in its
Three Homes" and "The Practical Bearings of
European History," which were afterward published
in a volume. After returning home he published
"Some Impressions of the United States" (1883),
a very entertaining book because of the author's
ingrained habit of comparing and discriminating
social phenomena upon so wide a scale. Gauls and
Illyrians, Wessex and Achaia, come in to point
each a moral, and show how to this great historian
the whole European past was almost as much a
present and living reality as the incidents occurring
before his eyes.

In the same year, 1883, Freeman published
his "English Towns and Districts," a series of
addresses and sketches in which he had from time
to time embodied the results of his antiquarian and
architectural studies in many parts of England and
Wales. It is a book of rare fascination as illustrating how largely national history is made up of

local history, and how it is impossible to understand the former correctly without paying much attention to the latter.   In further illustration of the same point, Freeman projected the well-known series of monographs on " Historic Towns," to which he himself contributed the opening volume, on " Exeter " (1886).

Having been called to the Regius Professorship at Oxford in 1884, Freeman's next publications were university lectures on " Methods of Historical Study," " The Chief Periods of European History," " Fifty Years of European History," " Teutonic Conquest in Gaul and Britain," " Greater Greece and Greater Britain," and " George Washington the Expander of England " (1886–88).   Meanwhile, the colossal work on " Sicily " was rapidly assuming its final shape.   This topic obviously touched upon Freeman's other two chief topics at two points. Ancient Sicily was part of that Greek world which he had so thoroughly studied in connection with the beginnings of federal government.   Mediæval Sicily was one of the most important of the Norman's fields of activity.   But the thought of writing the history of that fateful island did not come to Freeman as an afterthought suggested by his other two great works.   On the contrary, the conception of the historic position of Sicily was among the first that stimulated his philosophic mind to undertake comprehensive studies. The contact between the Aryan and Semitic civilizations along the coasts of the Mediterranean is surely the most interesting topic in the history of mankind, as the reader will at once admit when he reflects that it involves the origin and rise of Christianity. But, restricting ourselves to the political aspects of the subject, how full of dramatic grandeur it is! How stirring were the scenes of which Sicily has been the theatre! There struggled Carthage, first against Greek, and then against Roman; and in later times the conflict was renewed between Arabic-speaking Mussulmans and Greek-speaking Christians, until the Norman came to assert his sway over both, and to loosen the clutch of the Saracen upon the centre of the Mediterranean world. The theme, in its manifold bearings, was worthy of Freeman, and he was worthy of it. His design was to start with the earliest times in which Sicily is known to history, and to carry on the narrative as far as the death of the Emperor Frederick II. and the final overthrow of the Hohenstaufen dynasty. The scheme lay ripening in his mind for nearly half a century, and its consummation was begun with characteristic swiftness and vigour. Two noble volumes were published in 1891, and the third was out of the author's

hands by the end of last January. But for a
death most lamentably sudden and premature
there was no reason why the whole task should
not have been soon accomplished. The author
seems to have fallen a victim to his superabundant
zeal and energy. He had always been a traveller,
visiting in person the scenes of his narratives, nar-
rowly scrutinizing each locality with the eye of an
antiquarian, exploring battlefields and making drawings of churches and castles, running from one end
of Europe to the other to verify some mooted point.
It was, I believe, on some such expedition as this
that he found himself, last March, at Alicante,
where an attack of smallpox suddenly ended his
life.

To the faithful students of his works the tidings
of Freeman's death must have come like the news
of the loss of a personal friend. To those who
enjoyed his friendship even in a slight way the
sense of loss was keen, for he was a very lovable
man. Some people, indeed, seem to think of him
as a gruff and growling pedant, ever on the lookout for some culprit to chastise ; but, while not
without some basis, this notion is far from the
truth. Freeman's conception of the duty of a
historian was a high one, and he lived up to it.
He had a holy horror of slovenly and inaccurate

work ; pretentious sciolism was something that he could not endure, and he knew how easy it is to press garbled or misunderstood history into the service of corrupt politics. He found the minds of English-speaking contemporaries full of queer notions of European history, especially as to the Middle Ages, — notions usually misty and often grotesquely wrong; and he did more than any other Englishman of our time to correct such errors and clear up men's minds. Such work could not be done without attacking blunders and the propagators of blunders. Freeman's assaults were not infrequent, and they were apt to be crushing ; but they were made in the interests of historic truth, and there were none too many of them. Like " Mr. F.'s Aunt," the great historian did " hate a fool ; " and it is clearly right that fools should be silenced and made to know their place.

Not only foolishness and inaccuracy did Freeman hate, but also tyranny, fraud, and social injustice, under whatever specious disguises they might be veiled. In matters of right and wrong his perceptions were rarely clouded. He never could be duped into admiring a charlatan like the late Emperor of the French. Upon the Eastern Question he wielded a Varangian axe, and had his advice been heeded, the Commander of the Faith-

ful would ere now have been sent back to Brusa,
or beyond. But while in politics and in criticism
he could hit hard, his disposition was as tender and
humane as Uncle Toby's. Eminently characteristic is the discussion on fox-hunting which he carried on with Anthony Trollope some years ago in
the " Fortnightly Review," in which he condemned
that time-honoured sport as intolerably cruel.

Mr. Freeman was very domestic in his habits.
When not travelling, he was to be found in his
country home, writing in his own library. When
he was in the United States, it amused him to see
people's surprise when told that he did not live in
a city, and did not spend his time deciphering
musty manuscripts in public libraries or archives.
He used to say that, even in point of economy, he
thought it better to dwell among pleasant green
fields and to consult one's own books than to take
long journeys or be stifled in dirty cities in order
to consult other people's books. His chief subjects
of study favoured such a policy, for most of the
sources of information on the eleventh century,
as well as upon ancient Greece, are contained in
printed volumes. Now and then he missed some
little point upon which a manuscript might have
helped him. But one cannot help wishing he
might have stayed among the quiet fields of Somer-

set instead of taking that last fatal journey to Alicante.

It was chiefly with the political aspects of history that Freeman concerned himself; not in the old-fashioned way, as a mere narrative of the deeds of kings and cabinets, but in scientific fashion, as an application of the comparative method to the various processes of nation-building. I do not mean that his narrative was subordinated to scientific exposition, but that it was informed and vitalized by the spirit and methods of science. In pure description Freeman was often excellent; his account of the death of William Rufus, for example, is a masterpiece of impressive narrative. In description and in argument alike Freeman usually confined his attention to political history, except when he dealt in his suggestive way with architecture and archæology. To art in general, to the history of philosophy and of scientific ideas, to the development of literary expression, of manners and customs, of trade and the industrial arts, he devoted much less thought. I believe he did not fully approve of his friend Green's method of carrying along political, social, and literary topics abreast in his "History of the English People." Few will doubt, however, that in this respect Green's artistic grasp upon his subject was stronger than Freeman's.

It is some slight consolation for our bitter loss to know that many of the great historian's books were in large part written long before he felt the time to be ripe for completing and publishing them. Some of the unfinished portions may be brought toward completeness and edited by other hands. In this way I hope we may look for one or two more volumes of the " Sicily," and perhaps for the second volume of the " Federal Government," dealing with the Swiss and other German federations. Probably no other Englishman, and few other men anywhere in our time, knew anything like so much as Freeman about the history of Switzerland. I once or twice begged him to make haste and finish that volume, but desisted ; for it was evident that " Sicily " was absorbing him, and an author does not like to be pestered with advice to turn aside from the work that is uppermost in his mind.

November, 1892.

X

CAMBRIDGE AS VILLAGE AND CITY [1]

WE have met together this evening on one of those occasions, which keep recurring, for communities as well as for individuals, when it is desirable to take a retrospect of the past, to call attention to some of the characteristic incidents in our history, to sum up the work we have done and estimate the position we occupy in the world. As long as we retain the decimal numeration that is natural to ten-fingered creatures, we shall encounter such moments at intervals of half centuries and centuries, and happy are the communities that can meet them without shameful memories that shun the light of history; happy are the people that can look back upon the work of their fathers and in their heart of hearts pronounce it good! What a blot it was upon the civic fame of every Greek community that took part in putting out the

[1] An oration delivered in Sanders Theatre, June 2, 1896, at the civic jubilee commemorating the incorporation of Cambridge as a city.

brightest light of Hellas in the wicked Pelopon-
nesian War! Can any right-minded Venetian
look without blushing at the bronze horses that sur-
mount the stately portal of St. Mark's? — a perpetual memento of that black day when ravening
commercial jealousy decoyed an army of Crusaders
to the despoiling of the chief city of Christendom,
and thus broke away the strongest barrier in the
path of the advancing Turk! What must the
citizen of Paris think to-day of cowardly massacres
of unresisting prisoners, such as happened in 1418
and in 1792? Is there any dweller in Birmingham
who would not gladly expunge from the past that
summer evening which witnessed the burning of
the house and library of Dr. Priestley? From
such melancholy scenes, and from complicity in
political crime, our community, our neighbourhood,
has been notably free. The annals of Massachusetts, during its existence of nearly three centuries,
are written in a light that is sometimes dull or
sombre, but very seldom lurid. In particular the
career of Cambridge has been a placid one. We
do not find in it many things to startle us; but
there is much that we can approve, much upon
which, without falling into the self-satisfied mood
that is the surest index of narrowness and pro-
vincialism, we may legitimately pride ourselves.

In commemorating the fiftieth anniversary of the incorporation of Cambridge as a city, a retrospect of the half century is needful ; but we shall find it pleasant to go farther back, and start with a glimpse of the beginnings of our town.

I came near saying "humble beginnings ; " it is a stock phrase, and perhaps savours of tautology, since beginnings are apt to be humble as compared with long-matured results. But an adjective which better suits the beginnings of our Cambridge is "dignified." Circumstances of dignity attended the selection of this spot upon the bank of Charles River as the site of a town, and there was something peculiarly dignified in the circumstances of the change of vocation which determined the change in its name. The story is a very different one from that of the founding of towns in the Old World, in the semi-barbarous times when the art of nation-making was in its infancy. In those earlier ages, it was only through prolonged warfare against enemies nearly equal in prowess and resources that a free political life could be maintained ; and it was only after numberless crude experiments that nations could be formed in which political rights could be efficiently preserved for the people. All the training that such long ages of turbulence could impart had been gained by our forefathers in the Old World. To the founders of our Cambridge it had come as a rich inheritance. They were not as the rough followers of Alaric or Hengist. They had profited by the work of Roman civilization, with its vast and subtle nexus of legal and political ideas. In the hands of their fathers had been woven the wonderful fabric of English law; they were familiar with parliamentary institutions; they had been brought up in a country where the king's peace was better preserved than anywhere else in Europe, and where at the same time self-government was maintained in full vigour. They had profited, moreover, by the scholastic learning of the Middle Ages and the Greek scholarship of the Renaissance; nor was the newly awakening spirit of scientific inquiry, visible in Galileo and Gilbert, lost upon their keen and inquisitive minds. These Puritans, heirs to what was strongest and best in the world's culture, came to Massachusetts Bay in order to put into practice a theory of civil government in which the interests both of liberty and of godliness seemed to them likely to be best subserved. They came to plant the most advanced civilization in the midst of a heathen wilderness, and thus the selection of a seat of government for the new commonwealth was an affair of dignity and importance.

Half a dozen towns, including Boston, had already been begun, when it was decided that a site upon the bank of Charles River, three or four miles inland, would be most favourable for the capital of the Puritan colony. It would be somewhat more defensible against a fleet than the peninsulas of Boston and Charlestown. The warships to be dreaded at that moment were not so much those of any foreign power as those of King Charles himself; for none could tell that the grim clouds of civil war then lowering upon the horizon of England and Scotland might not also darken the coast of Massachusetts Bay. When the site was selected, on the 28th of December, 1630, it was agreed that the governor, deputy governor, and all the Court of Assistants (except Endicott, already settled at Salem) should build their houses here. Fortunately no name was bestowed upon the new town. It was known simply as the New Town, and here in the years before 1638 the General Court was several times assembled. During those seven years the number of Puritans in New England increased from about 1500 to nearly 20,000. It was also clear that the King's troubles at home were likely to keep him from molesting Massachusetts. With the increased feeling of security, Boston came to be preferred as the seat of government,

and only two of its members ever fulfilled the
agreement to build their houses in the New Town.

The building of the New Town, however, furnished the occasion for determining at the outset
what kind of government the Puritan commonwealth should have. It was to be a walled town,
for defence against frontier barbarism of the New
World type; not the formidable destructive power
of an Attila or a Bayazet, but the feeble barbarism
of the red men and the Stone Age, so that a wall
of masonry was not required, but a wooden palisade
would do. In 1632 the Court of Assistants im-
posed a tax of £60 for the purpose of building this
palisade; but the men of Watertown refused to
pay their share, on the ground that they were not
represented in the taxing body. The ensuing discussion resulted in the establishment of a House of
Deputies, in which every town was represented.
Henceforth the Court of Assistants together with
the House of Deputies formed the General Court.
There was no authority for such a representative
body in the charter, which vested the government
in the Court of Assistants; but, as Hutchinson
tells us, the people assumed that the right to such
representation was implied in that clause of the
charter which reserved to them the natural rights
of Englishmen. Thus the building of a wooden

palisade from Ash Street to Jarvis Field furnished the occasion for the first distinct assertion in the New World of the principles that were to bear fruit in the independence of the United States.

But the most interesting event in the history of the New Town before it became Cambridge was the brief sojourn of the Rev. Thomas Hooker and his company, from Braintree in England. In popular generalizations it is customary to allude to our Puritan forefathers as if they were all alike in their ways of thinking, whereas in reality it would be difficult to point out any group of men and women among whom individualism has more strongly flourished. Among the numberless differences of opinion and policy, it was only a few — and mostly such as were related to vital political questions — that blazed up in acts of persecution. For the disorganization wrought by Mrs. Hutchinson swift banishment seemed the only available remedy; but slighter differences could be healed by a peaceful secession, which some people deprecated as the " removal of a candlestick." Such a secession was that of Hooker and his friends. The difference between Hooker's ideal of government and Winthrop's has come to be recognized as in some measure foreshadowing the different conceptions of Jefferson and Hamilton in later days. But of controversy between the two eminent Puritans only slight traces are left. One act of omission on the part of the friendly seceders is more forcible than reams of argument: the founders of Connecticut did not see fit to limit the suffrage by the qualification of church membership.

The removal of so many people to the banks of the Connecticut left in the New Town only eleven families of those who had settled here before 1635. But depopulation was prevented by the arrival of a new congregation from England. There stands on our common a monument in commemoration of John Bridge, who was for many years a selectman of Cambridge, and dwelt beyond the western limits of the town, on or near the site since famous as the headquarters of Washington and the home of Longfellow. This John Bridge, deacon of the First Church, was one of the earliest settlers of the New Town, and one of the eleven householders that stayed behind, a connecting link between the old congregation of Thomas Hooker and the new congregation of Thomas Shepard. The coming of this eminent divine was undoubtedly an event of cardinal importance in the history of our community, for in the Hutchinson controversy, which shook the little colony to its foundations, his zeal

and vigilance in exposing heresy were conspicuously shown; and, if we may believe Cotton
Mather, it was this circumstance that led to the
selection of the New Town as the site for the projected college. It was well for students of divinity
to sit under the preaching of such a man, and of
such as he might train up to succeed him. How
vain were all such hopes of keeping this New
English Canaan free from heresy was shown when
Henry Dunster, first president of the college, was
censured by the magistrates and dismissed from
office for disapproving of infant baptism!

In the great English universities at that time
Royalism and Episcopacy prevailed at Oxford,
while Puritanism more or less allied with Repub-
licanism was rife at Cambridge. Ever since the
fourteenth century a superior flexibility in opinion had been observable in the eastern counties,
whence came so many of the people that founded
New England. Not only Hooker and Shepard,
but most of our clergy, among whom individualism
was so rife, were graduates of Cambridge. When
it was decided that the New Town was to be the
home of our college, it was natural for people to
fall into the habit of calling it Cambridge; and
this name, so long enshrined already in their affec-
tions, already made illustrious by Erasmus and

Fisher, by Latimer and Cranmer, by Burghley and Walsingham and the two Bacons, by Edmund Spenser and Ben Jonson, — this name of such fame and dignity was adopted in 1638 by an order of the General Court. The map of the United States abounds in town names taken at random from the Old World, often inappropriate and sometimes ludicrous from the incongruity of associations. The name of our city is connected by a legitimate bond of inheritance with that of the beautiful city on the Cam. It was given in the thought that the work for scholarship, for godliness, and for freedom, which had so long been carried on in the older city, was to be continued in the younger. The name thus given was a pledge to posterity, and it has been worthily fulfilled.

Into the history of the town of Cambridge during the two centuries after it received its name I do not propose to enter. But a glimpse of its general appearance during the greater part of that period is needful, in order to give precision and the right sort of emphasis to the contrast which we see before us to-day. The Cambridge of those days was simply the seat of the college, not yet developed into a university. Within the memory of persons now living, Old Cambridge was commonly alluded to as " the village." In the original laying

out of the township we seem to see a reminiscence
of the ancient threefold partition into town mark,
arable mark, and common. The " east gate," near
the corner of Harvard and Linden streets, and the
" west gate," at the corner of Ash and Brattle,
marked the limits of the town in those directions.
The town was at first comprised between Harvard
Street and the marshes which cut off approach
to the river bank. Afterward, the " West End,"
from Harvard Square to Sparks Street, was gradually covered with homesteads. The common began, as now, hard by God's Acre, the venerable
burying ground, and afforded pasturage for the
village cattle as far as Linnæan Street. The regions now occupied by Cambridgeport and East
Cambridge contained the arable district with many
farms, small and large, but everywhere salt
marshes bordered the river, and much of the country was a wild woodland. The tale of wolves killed
in Cambridge for the year 1696 was seventy-six,
and a bear was seen roaming as late as 1754. It
was a rough country which the British first encountered when they landed at Lechmere Point in 1775,
on their night march to Lexington. Cambridge
then turned its back toward Boston, to which the
only approach was by a causeway and bridge at
what we now call Boylston Street, and by this route

the distance was eight miles, as we still read upon
the ancient milestone in God's Acre. To complete
our outline of the village, we must recall the principal public buildings. The meetinghouse, a little
south of the site of Dane Hall, was used both as
church and as townhouse until 1708, when a building was erected in the middle of Harvard Square
to serve for town meetings and courts. A little
eastward, near the " east gate," stood the parson-
age. The schoolhouse was behind the site of Hol-
yoke House. The jail stood on the west side of
Winthrop Square, which was then an open market.
Between this market and Harvard Square, in the
sanded parlour of the Blue Anchor Tavern, the
selectmen held their meetings ; and on the corner
of the street which still bears the name of Harvard's first president was something rarely to be
seen in so small a village, the printing press now
known as the University Press, established in
1639, — the only one in English America until
Boston followed the example in 1676.

Until the beginning of the present century these
outlines of Cambridge remained with but little
change, save for the building of noble houses on
spacious estates toward Mount Auburn in one direction, and upon Dana Hill in the other. The oc-
cupants of many of these estates were members of

the Church of England, and the building of Christ
Church in 1759 was one marked symptom of the
change that was creeping over the little Puritan
community. It was a change toward somewhat
wider views of life, and toward the softening of
old animosities. In contrast with the age in which
we live the whole eighteenth century in New England seems a slow and quiet time, when the public
pulse beat more languidly, or at any rate less fe-
verishly, than now. The people of New England
led a comparatively isolated life.

Thought in our college town did not keep pace
with European centres of thought, as it does in our
day. There was less hospitality toward foreign
ideas. Few people visited Europe. Life in New
England was thrown upon its own resources, and
this was in great measure true of Cambridge in the
days when it was eight miles from Boston, and in-
definitely remote from the mother country. One
of the surest results of social isolation is the acquirement of peculiarities of speech, often shown
in the retaining of archaisms which fashionable
language has dropped. That quaint Yankee dialect, of which Hosea Biglow says that,

> " For puttin' in a downright lick
>   'Twixt Humbug's eyes, ther 's few can metch it,
>   An' then it helves my thoughts ez slick
>   Ez stret-grained hickory doos a hetchet," —

that dialect so sweet to the ears of every true child
of New England may still be heard, if we go to
seek it ; but in Lowell's boyhood it must have been
a familiar sound in the neighbourhood of Elmwood.

But the work done in this rustic college community, if done within somewhat narrow horizons,
was eminently a widening and liberalizing work.
The seeds of the nineteenth century were germinating in the eighteenth. Two or three indications
must suffice, out of many that might be cited.   In
1669 there was a schism in the First Parish of
Boston, brought about by an attempt to revise the
conditions of church membership, in order to obvi-
ate some of the difficulties arising from the restric-
tion of the suffrage to church members, and the
founding of the Old South Church by the more
liberal party was a result of this schism.   One
hundred and sixty years later, in 1829, there was
a schism in the First Parish of Cambridge, which
resulted in the founding of the Shepard Church
by the more conservative party.   The questions at
issue between the two parties were the questions
that divide Unitarian theology from Trinitarian,
and the distance between the kind of interests at
stake in the earlier controversy and in the later
may serve as a fair measure of the progress which
the mind of Massachusetts had been making during

that interval of a hundred and sixty years. In all
that time, the chief training school for the ministers by whom the speculative minds of Massachusetts were stimulated and guided was Harvard College. But it was here, too, that men eminent in
civic life were trained ; and among the various
illustrations of the type thus nurtured may be cited
Samuel Adams and Thomas Hutchinson, foemen
worthy of one another, Warren and Hancock, Jonathan Trumbull and John Adams. So far as New
England was concerned, the chief work in bringing
on the Revolution was done by graduates of Harvard. In the convention which framed our Federal Constitution, three important delegates were
the Harvard men, Gerry, Strong, and King; and
in this connection we cannot fail to recall names
so closely associated with our national beginnings
as Timothy Pickering and Fisher Ames, nor can
we omit the noble line of jurists from Parsons to
Story, and so on to Curtis, whom so many of us
well remember ; or, going back to that Massachusetts convention, of which the work is commemorated in the name of Federal Street, we may single
out for mention the great minister and statesman,
type of what is best in Puritanism, Samuel West,
of New Bedford. Such names speak for the kind
of quiet, unobtrusive work that was going on in

Cambridge during those two centuries of rural existence. Such strengthening and unfolding of the
spirit is the only work that is truly immortal. In
a town like ours the material relics of the past are
inspiring, and it is right that we should do our best
to preserve them ; but they are perishable. The
gambrel-roofed house from the door of which President Langdon asked God's blessing upon the men
that were starting for Bunker Hill, in later days
the birthplace and homestead of our beloved Autocrat, has vanished from the scene ; the venerable
elm under which Washington drew the sword in
defence of American liberty is slowly dying, year
by year. But for the spiritual achievement that
has marked the career of our community there is
no death, and they that have turned many to righteousness shall shine in our firmament as the stars
forever and ever.

In contrasting the Cambridge of the nineteenth
with that of the two preceding centuries, the first
fact which strikes our attention is the increase in
the rate of growth. In 1680 the population of
Cambridge seems to have been about 850, and the
graduating class for that year numbered five. In
1793 the population — not counting the parishes
that have since become Brighton and Arlington —
was about 1200, and there was a graduating class

of 38. Thus in more than a century the population had increased barely fifty per cent. In 1793 there were only four houses east of Dana Street, but that year witnessed an event of cardinal importance, the opening of West Boston Bridge. The distance between Boston and Old Cambridge was thus reduced from eight miles to three, and a direct avenue was opened between the interior of Middlesex County and the Boston markets. The effect was shown in doubling the population of Cambridge by the year 1809, when another bridge was completed from Lechmere Point to the north end of Boston. These were toll bridges, in the hands of private corporations, and their success led to further bridges, — the one at River Street in 1811, the one at Western Avenue in 1825, and Brookline Bridge so lately as 1850. The principal thoroughfares south and east of Old Cambridge were built as highways connecting with these bridges: thus River Street and Western Avenue were tributary to West Boston Bridge, and to that point the Concord Turnpike was prolonged by Broadway, the Middlesex Turnpike by Hampshire Street, and the Medford Road by Webster Avenue; while Cambridge Street, intersecting these avenues, formed a direct thoroughfare from the Concord and Watertown roads to the northern

part of Boston. The completion of these important works led to projects for filling up the marshes
and establishing docks in rivalry of Boston, —
plans but very slightly realized before circumstances essentially changed them.

In this way, Cambridge, which had hitherto faced
the Brighton mainland, turned its face toward the
Boston peninsula, and two new villages began to
grow up at " the Port " and " the Point," otherwise
Cambridgeport and East Cambridge. It was not
long before the new villages began in some ways to
assert rivalry with the old one. The corporation
which owned the bridge and large tracts of land at
Lechmere Point naturally wished to increase the
value of its real estate. Middlesex County needed
a new courthouse and jail. In 1757 a new courthouse had been built on the site of Lyceum Hall,
but in 1813 there was a need for something better ;
whereupon the Lechmere Point Corporation forthwith built a courthouse and jail in East Cambridge, and presented them, with the ground on
which they stood, to the county. In 1818, a lot
of land in the Port, bounded by Harvard, Prospect, Austin, and Norfolk streets, was appropriated
for a poorhouse. Soon afterward it was proposed
to inclose our common, — which with the lapse of
time had shrunk to about its present size, — and

to convert it from a grazing ground into an ornamental park. The scheme met with vehement opposition, and the town meetings in this growing community suddenly became so large that the old courthouse in Harvard Square would not hold them. Accordingly, a bigger townhouse was built in 1832 on the eastern part of the poorhouse lot, and thus was the civic centre removed from Old Cambridge.

This event served to emphasize the state of things which had been growing up with increasing rapidity since the beginning of the century. Instead of a single village, with a single circle of interests, there were now three villages, with interests diverse and sometimes conflicting as regards the expending of public money, so that feelings of sectional antagonism were developed.

In New England history, the usual remedy for such a state of things has been what might be called "spontaneous fission." The overgrown town would divide into three, and the segments would go on pouting at each other as independent neighbours. We need not be surprised to learn that in 1842 the people of Old Cambridge petitioned to be set off as a separate town; but this attempt was successfully opposed, with the result that in 1846 a city government was adopted. In that year the population had reached 13,000, and was approaching the point at which town meetings become unmanageable from sheer bulk. For small communities, Thomas Jefferson was probably right in holding that the town meeting is the best form of government ever devised by man. It was certainly the form best loved in New England down to 1822, when Boston, with its population of 40,000, reluctantly gave it up, and adopted a representative government instead. The example of Boston was followed in 1836 by Salem and Lowell, and next in 1846 by Roxbury and Cambridge. From that time forth the making of cities went on more rapidly. It was the beginning of a period of urban development, the end of which we cannot as yet even dimly foresee. This unprecedented growth of cities is sometimes spoken of as peculiarly American, but it is indeed not less remarkable in Europe, and it extends over the world so far as the influence of railroad and telegraph extends. The influence of these agencies of communication serves to diffuse over wide areas the effects wrought by machinery at different centres of production. With increased demand for human energy, the earth's power of sustaining human life has vastly increased, and there is a strong tendency to congregate about centres of production and exchange. In 1846 there were but five cities in the United States with a population exceeding 100,000 ; New York had not yet reached half a million. To-day New York is approaching the two-million mark, three other cities [1] have passed the million, and not less than thirty have passed the hundred-thousand. During this half century the 13,000 of Cambridge have increased to more than 80,000. The Cambridge of to-day contains as many people as the Boston of sixty years ago.

The causes of this growth of Cambridge might be treated, had we space for it, under three heads. Our city has grown because of its proximity to Boston ; it has grown by reason of its flourishing manufactures ; and it has grown with the growth of the University. That Cambridge should have shared in the general prosperity of this whole suburban region is but natural. But persons at a distance are apt to show surprise when we speak of it as a manufacturing city. This feature in our development belongs to the period subsequent to 1846, and has much to do with the growth of the eastern portions of Cambridge, where the combined facilities for railroad and water communication have

---

[1] Chicago, Philadelphia, and Brooklyn. By the annexation of Brooklyn, the population of New York is now (1899) carried up to 3,500,000, making it the second city in the world.

been peculiarly favourable to manufactures. In
the early part of this century, the glassworks at
East Cambridge, which have since departed, were
somewhat famous, considerable manufactures of
soap and leather had been begun, and cars and
wagons were made here. At the present time
some of our chief manufactures are of engine
boilers and various kinds of machinery, of which
the annual product exceeds $2,000,000. Among
the industries which produce in yearly value more
than $1,000,000 may be mentioned printing and
publishing, musical instruments (especially pianos
and organs), furniture, clothing, carpenter's work,
soap and candles, biscuit-baking ; while among
those that produce $500,000 or more are carriage-
making and wheelwright's work, plumbing and
plumber's materials, bricks and tiles, and confec-
tionery. Not only our own new Harvard Bridge,
but most of the steel railway bridges in New England, have been built in Cambridge. We supply
a considerable part of the world with hydraulic engines ; the United States Navy comes here for its
pumps, and our pumping machines may be seen at
work in Honolulu, in Sydney, in St. Petersburg.
In the dimensions of its pork-packing industry,
Cambridge comes next after Chicago and Kansas
City. In 1842 all the fish-netting used in Amer-

ica was made in England; to-day it is chiefly
made in East Cambridge, which also furnishes the
twine prized by disciples of Izaak Walton in many
parts of the world. Last year the potteries on
Walden Street turned out seven million flower-
pots. Such facts as these bear witness to the unusual facilities of our city, where coal can be taken
and freight can be shipped at the very door of the
factory, where taxes and insurance are not burden-
some and the fire department is unsurpassed for
efficiency, where skilled labour is easy to get because
good workmen find life comfortable and attractive,
with excellent sanitary conditions and unrivalled
means of free education, even to the Latin School
and the Manual Training School. It is well said,
in one of the reports in our semi-centennial volume, that " to Cambridge herself, as much as to
any other one thing, is the success of all her manufacturing enterprises due, and all agree in acknow-
ledging it."

Among Cambridge industries, two may be mentioned as especially characteristic and famous. Of
the printing establishments now existing, not many
can be more venerable than our University Press,
of which we have spoken as beginning in 1639.
Of the wise and genial founder of the Riverside
Press — who once was mayor of our city, and

whose memory we love and revere — it may be said that few men of recent times have had a higher conception of bookmaking as one of the fine arts. These two institutions have set a lofty standard for the Athenæum Press, which has lately come to bear them company. The past half century has seen Cambridge come into the foremost rank among the few publishing centres of the world, where books are printed with faultless accuracy and artistic taste.

The visitor to Cambridge from Brookline, as he leaves the bridge at Brookline Street, comes upon a pleasant dwelling house, with a private observatory, and hard by it a plain brick building. That is the shop of Alvan Clark & Sons, who have carried the art of telescope-making to a height never reached before. There have been made the most powerful refracting telescopes in the world, and one of the firm, more than thirty years ago, himself acquired fame as an astronomer for his discovery of the companion of Sirius.

From this quiet nook in the Port one's thoughts naturally turn to the Harvard Observatory, which in those days the two Bonds made famous for their accurate methods of research, their discoveries relating to the planet Saturn, and their share in the application of photography to telescopic

observation. The honourable position then taken
by the Observatory has been since maintained;
but as we note this, we find ourselves brought to
the consideration of the University and its last
half century of growth. And here my remarks
cannot help taking the form, to some extent, of
personal reminiscences.

When I first came to Old Cambridge, in 1860,
it still had much of the village look, which it has
since been fast losing. Pretty much all the spaces
now covered by street after street of wooden " Queen
Anne" houses, in such proximity as to make one
instinctively look for the whereabouts of the nearest
fire alarm, were then open, smiling fields. The old
house where the Shepard Church stands was rural
enough for the Berkshire Hills; and on the site of
Austin Hall, in the doorway of a homestead built
in 1710, one might pause for a cosy chat with the
venerable and courtly Royal Morse, whose personal
recollections went back into the eighteenth century.
The trees on the common were the merest sap-
lings, but an elm of mighty sweep, whose loss one
must regret, shaded the whole of Harvard Square.
Horse cars came and went on week days, but on
Sunday he who would visit Boston must either
walk or take an omnibus, in which riding was a
penance severe enough to atone for the sin. " Blue

Laws " in the University were in full force ; the
student who spent his Sundays at home in Boston
must bring out a certificate showing that he had
attended divine service twice ; no discretion was
allowed the parents.

College athletics were in their infancy, as the
little gymnasium still standing serves to remind
us.   There were rowing matches, but baseball had
not come upon the scene, and football had just
been summarily suppressed.   The first college
exercise in which I took part was the burial of the
football, with solemn rites, in a corner of this
Delta.   On Class Day there was no need for clos-
ing the yard ; there was room enough for all, and
groups of youths and maidens in light summer
dress, dancing on the green before Holworthy,
made a charming picture, like that of an ancient
May Day in merry England, save for the broiling
heat.

The examination days which followed were more
searching than at other American colleges.   The
courses of study were on the whole better arranged
than elsewhere, but during the first half of the
course everything was prescribed, and in the last
half the elective system played but a subordinate
part.   The system of examinations did not extend
to the Law School, where a simple residence of

three terms entitled a student to receive the bachelor's degree. The library at Gore Hall had less than one fifth of its present volumes, with no catalogue accessible to the public, while one small table accommodated all the readers. For laboratory work the facilities were meagre, and very little was done. We all studied a book of chemistry; how many of us ever really looked at such things as manganese or antimony? For the student of biology the provision was better, for the Botanic Garden was very helpful, and in the autumn of 1860 was opened the first section of our glorious Museum of Comparative Zoölogy.

Here one is naturally led to the reflection that in that day of small things, as some might call it, there were spiritual influences operative at Harvard which more than made up for shortcomings in material equipment. There is a kind of human presence, all too rare in this world, which is in itself a stimulus and an education worth more than all the scholastic artifices that the wit of man has devised; for in the mere contact with it one's mind is trained and widened as if by enchantment. Such a human presence in Cambridge was Louis Agassiz. Can one ever forget that beaming face as he used to come strolling across the yard, with lighted cigar, in serene obliviousness of the University statutes? Scarcely had one passed him, when one might exchange a pleasant word with Asa Gray, or descry in some arching vista the picturesque figures of Sophocles or Peirce, or, turning up Brattle Street, encounter, with a thrill of pleasure not untinged with awe, Longfellow and Lowell walking side by side. In such wise are the streets and lawns of our city hallowed by the human presences that once graced them; and few are the things to be had for which one would exchange the memories of those days!

My class of 1863, with 120 members, was the largest that had been graduated here. It would have been larger but for the Civil War, and a period followed with classes of less than 100 members, — a sad commentary upon the times. Boundless possibilities of valuable achievement must be sacrificed to secure the supreme end, that the commonwealth should not suffer harm. How nobly Harvard responded to the demand is recorded upon the solemn tablets in this Memorial Hall. For those who are inclined to dally with the thought that war is something that may be undertaken lightly and with frolicsome heart, this sacred precinct and the monument on yonder common have their lesson that may well be pondered.

The vast growth of our country since the Civil

War has been attended with the creation of new universities and the enlargement of the old ones to such an extent as to show that the demand for higher education more than keeps pace with the increase of population. The last graduating class in our Quinquennial Catalogue numbered 350 members. The University contains more than 3000 students.[1] The increase in number of instructors, in courses of instruction, in laboratories and museums, in facilities and appliances of every sort, has wrought changes like those in a fairy tale. The Annual Catalogue is getting to be as multifarious as Bradshaw's Guide, and a trained intellect is required to read it. The little college of half a century ago has bloomed forth as one of the world's foremost universities. Such things can come from great opportunities wielded and made the most of by clearness of vision and administrative capacity.

To this growth of the University must be added the most happy inception and growth of Radcliffe College, marking as it does the maturing of a new era in the education of women. We may well wish for Radcliffe a career as noble and as useful as that of Harvard, and I doubt not that such is in store

---

[1] In 1898 the number had risen to 4660, besides 411 women students in Radcliffe.

for it. A word must be said of the Episcopal Theological School, based upon ideas as sound and broad as Christianity; and of the New-Church Theological School, more recently founded. We must hail such indications of the tendency toward making our Cambridge the centre for the untrammelled study of the most vital problems that can occupy the human mind.

But the day we are celebrating is a civic, not a university occasion, and I must dwell no longer upon academic themes. We are signalizing the anniversary of the change which we once made from government by town meeting to city government. Have we a good reason for celebrating that change? Has our career as a civic community been worthy of approval? In answering this question, I shall not undertake to sum up the story of our public schools and library; our hospital and charity organizations; the excellent and harmonious work of our churches, Protestant and Catholic; our Prospect Union, warmly to be commended; our arrangements for water supply and sewage; and our admirable park system (in which we may express a hope that Elmwood will be included). This interesting and suggestive story may be read in the semi-centennial volume, "The Cambridge of Eighteen Hundred and Ninety-Six," just issued

from the Riverside Press. It is an enlivening
story of progress, but like every story it has a
moral, and I am going to pass over details and
make straight for that moral. Americans are a
bragging race because they have enjoyed immense
opportunities, and are apt to forget that the true
merit lies, not in the opportunity, but in the use
we make of it. Much gratifying progress can be
achieved in spite of the worst sort of blundering
and sinning on the part of governments. The
greater part, indeed, of human progress within historic times has been thus achieved. A good deal
of the progress of which Americans are wont to
boast has been thus achieved. Now the moral of
our story is closely concerned with the fact that in
the city of Cambridge such has not been the case.
Our city government has from the outset been upright, intelligent, and helpful. We are satisfied
with it. We do not wish to change it. In this
respect the experience of Cambridge is very different from that of many other American cities.
The government of our cities is acknowledged to
be a problem of rare difficulty, so that it has begun
to seem a natural line of promotion for a successful
mayor to elect him governor, and then to send him
to the White House! In some cities one finds
people inclined to give up the problem as insoluble.

I was lately assured by a gentleman in a city which
I will not name, but more than a thousand miles
from here, that the only cure for the accumulated
wrongs of that community would be an occasional
coup d'état, with the massacre of all the city offi-
cers.   So the last word of our boasted progress,
when it comes to municipal government, is declared
to be the Oriental idea of " despotism tempered by
assassination " !   Now to what cause or causes are
we to ascribe the contrast between Cambridge and
the cities that are so wretchedly governed ?   The
answer is, that in Cambridge we keep city government clear of politics, we do not mix up municipal
questions with national questions.   If I may repeat
what I have said elsewhere, "since the object of
a municipal election is simply to secure an upright and efficient municipal government, to elect
a city magistrate because he is a Republican or a
Democrat is about as sensible as to elect him because he believes in homœopathy or has a taste for
chrysanthemums."   Upon this plain and obvious
principle of common sense our city has acted, on
the whole with remarkable success, during its half
century of municipal existence.   The results we
see all about us, and the example may be commended as an object lesson to all who are interested in the most vital work that can occupy the

mind of an American, — the work of elevating the moral tone of public life. For it is neither wealth, nor power, nor cunning, nor craft that exalts a nation, but righteousness and the fear of the Lord.

May, 1896.

XI

A HARVEST OF IRISH FOLK-LORE

SINCE the days when Castrèn made his arduous journeys of linguistic exploration in Siberia, or when the brothers Grimm collected their rich treasures of folk-lore from the lips of German peasants, an active quest of vocables and myths has been conducted with much zeal and energy in nearly all parts of the world. We have tales, proverbs, fragments of verse, superstitious beliefs and usages, from Greenland, from the southern Pacific, from the mountaineers of Thibet and the freedmen upon Georgia plantations. We follow astute Reynard to the land of the Hottentots, and find the ubiquitous Jack planting his beanstalk among the Dog-Rib Indians. At the same time, the nooks and corners of Europe have been ransacked with bountiful results; so that whereas our grandfathers, in speculating about the opinions and mental habits of people in low stages of culture, were dealing with a subject about which they knew almost nothing, on the other hand, our chief difficulty to-day is in shaping and managing the enormous mass of data which keen and patient inquirers have collected. It is well that this work has been carried so far in our time, for modern habits of thought are fast exterminating the Old World fancies. Railroad, newspaper, and telegraphic bulletin of prices are carrying everything before them. The peasant's quaint dialect and his fascinating myth tales are disappearing along with his picturesque dress; and savages, such of them as do not succumb to fire-water, are fast taking on the airs and manners of civilized folk. It is high time to be gathering in all the primitive lore we can find, before the men and women in whose minds it is still a living reality have all passed from the scene.

The collection of Irish myth stories lately published by Mr. Jeremiah Curtin [1] is the result of a myth-hunting visit which the author made in Ireland in 1887, and is one of the most interesting and valuable contributions to the study of folk-lore that have been made for many years. "All the tales in my collection," says Mr. Curtin, "of which those printed in this volume form but a part, were taken down from the mouths of men who, with one or two exceptions, spoke only Gaelic, or but little

[1] *Myths and Folk-Lore of Ireland.* By Jeremiah Curtin. Boston: Little, Brown & Co. 1890.

English, and that imperfectly. These men belong
to a group of persons all of whom are well advanced in years, and some very old; with them
will pass away the majority of the story-tellers of
Ireland, unless new interest in the ancient language and lore of the country is roused.

"For years previous to my visit of 1887 I was
not without hope of finding some myth tales in a
good state of preservation. I was led to entertain
this hope by indications in the few Irish stories
already published, and by certain tales and beliefs
that I had taken down myself from old Irish persons in the United States. Still, during the earlier part of my visit in Ireland, I was greatly
afraid that the best myth materials had perished.
Inquiries as to who might be in possession of these
old stories seemed fruitless for a considerable time.
The persons whom I met that were capable of
reading the Gaelic language had never collected
stories, and could refer only in a general way to
the districts in which the ancient language was
still living. All that was left was to seek out the
old people for whom Gaelic is the every-day speech,
and trust to fortune to find the story-tellers."

Thus Mr. Curtin was led to explore the counties
of Kerry, Galway, and Donegal. "Comforting
myself with the Russian proverb that 'game runs

to meet the hunter,' I set out on my pilgrimage, giving more prominence to the study and investigation of Gaelic, which, though one of the two objects of my visit, was not the first. In this way I thought to come more surely upon men who had myth tales in their minds than if I went directly seeking for them. I was not disappointed, for in all my journeyings I did not meet a single person who knew a myth tale or an old story who was not fond of Gaelic, and specially expert in the use of it, while I found very few story-tellers from whom a myth tale could be obtained unless in the Gaelic language; and in no case have I found a story in the possession of a man or woman who knew only English."

There is something so interesting in this fact, and so pathetic in the explanation of it, that we are tempted to quote further: "Since all mental training in Ireland is directed by powers both foreign and hostile to everything Gaelic, the moment a man leaves the sphere of that class which uses Gaelic as an every-day language, and which clings to the ancient ideas of the people, everything which he left behind seems to him valueless, senseless, and vulgar; consequently he takes no care to retain it, either in whole or in part. Hence the clean sweep of myth tales in one part of the country, —

the greater part, occupied by a majority of the
people; while they are still preserved in other and
remoter districts, inhabited by men who, for the
scholar and the student of mankind, are by far the
most interesting in Ireland."

The fate of the Gaelic language has, indeed,
been peculiarly sad. In various parts of Europe,
and especially among the western Slavs, the native
tongues have been to some extent displaced by the
speech of conquering peoples; yet it is only in
Erin that, within modern times, a " language of
Aryan stock has been driven first from public use,
and then dropped from the worship of God and the
life of the fireside."   Hence, while in many parts
of Europe the ancient tales live on, often with their
incidents more or less dislocated and their significance quite blurred, on the other hand, in English-
speaking Ireland they have been cleared away " as
a forest is felled by the axe."

Nevertheless, in the regions where Irish myths
have been preserved, they have been remarkably
well preserved, and bear unmistakable marks of
their vast antiquity. One very noticeable feature
in these myths is the definiteness and precision of
detail with which the personages and their fields of
action are brought before us. This is a characteristic of mythologies which are, comparatively speak-

ing, intact; and, as Mr. Curtin observes, it is to be seen in the myths of the American Indians. As long as a mythology remains intact it "puts its imprint on the whole region to which it belongs." Every rock, every spring, is the scene of some definite incident; every hill has its mythical people, who are as real to the narrators as the flesh-and-blood population which one finds there. In this whole world of belief and sentiment there is the vigour of fresh life, and the country is literally enchanted ground. But when, through the invasion of alien peoples, there is a mingling and conflict of sacred stories, and new groups of ideas and associations have partly displaced the old ones, so that only the argument or general statement of the ancient myth is retained, and perhaps even that but partially, then "all precision and details with reference to persons and places vanish; they become indefinite; are in some kingdom, some place, — nowhere in particular." There is this vagueness in the folk-tales of eastern and central Europe as contrasted with those of Ireland. "Where there was or where there was not," says the Magyar, "there was in the world;" or, if the Russian hero goes anywhere, it is simply across forty-nine kingdoms, etc.; "but in the Irish tales he is always a person of known condition in a specified place" (for example, "There was a blacksmith in Dun-kenealy, beyond Killybegs," etc., page 244).

As to the antiquity and the primitive character of Mr. Curtin's stories an experienced observer can entertain no doubt. His book is certainly the most considerable achievement in the field of Gaelic mythology since the publication, thirty years ago, of Campbell's "Tales of the West Highlands;" and it does for the folk-lore of Ireland what Asbjörnsen and Moe's collection (the English translation of which is commonly, and with some injustice, known by the name of the translator as Dasent's "Norse Tales") did for the folk-lore of Norway. This is, of course, very high praise, but we do not believe it will be called extravagant by any competent scholar who reads Mr. Curtin's book. The stories have evidently been reduced to writing with most scrupulous and loving fidelity. In turning the Gaelic into English some of the characteristic Hibernian phrases and constructions of our language have been employed, and this has been done with such perfect good taste that the effect upon the ear is like that of a refined and delicate brogue.

The mythical material in the stories is largely that with which the student of Aryan folk-lore is familiar. We have variants of Cinderella, the

swan maidens, the giant who had no heart in his
body, the cloak of darkness, the sword of light, the
magic steed which overtakes the wind before and
outstrips the wind behind; the pot of plenty, from
which one may eat forever, and the cup that is
never drained; the hero who performs impossible
tasks, and wooes maidens whose beauty hardly re-
lieves their treacherous cruelty : " I must tell you
now that three hundred king's sons, lacking one,
have come to ask for my daughter, and in the
garden behind my castle are three hundred iron
spikes, and every spike of them but one is covered
with the head of a king's son who could n't do
what my daughter wanted of him, and I 'm greatly
in dread that your own head will be put on the one
spike that is left uncovered." The princess in this
story — Shaking - Head — is such a wretch, not
a bit better than Queen Labe in the " Arabian
Nights," that one marvels at the hero for marrying
her at last, instead of slicing off her head with his
two-handed sword of darkness, and placing it on
the three-hundredth spike. But moral as well as
physical probabilities are often overstrained in this
deliciously riotous realm of folk-lore.

Along with much material that is common to
the Aryan world there is some that is peculiar to
Ireland, while the Irish atmosphere is over every-

thing. The stories of Fin MacCumhail (pronounced MacCool) and the Fenians of Erin are
full of grotesque incident and inimitable drollery.
Fin and his redoubtable dog Bran, the one-eyed
Gruagach, the hero Diarmuid, the old hag with the
life-giving ointment, the weird hand of Mal Mac-
Mulcan, and the cowherd that was son of the king
of Alban make a charming series of pictures.
Among Fin's followers there is a certain Conán
Maol, " who never had a good word in his mouth
for any man," and for whom no man had a good
word. This counterpart of Thersites, as Mr. Curtin tells us, figures as conspicuously in North
American as in Aryan myths. Conán was always
at Fin's side, and advising him to mischief. Once
it had like to have gone hard with Conán. The
Fenians had been inveigled into an enchanted
castle, and could not rise from their chairs till two
of Fin's sons had gone and beheaded three kings
in the north of Erin, and put their blood into
three goblets, and come back and rubbed the blood
on the chairs. Conán had no chair, but was sitting on the floor, with his back to the wall, and
just before they came to him the last drop of blood
gave out. The Fenians were hurrying past without
minding the mischief-maker, when, upon his earnest appeal, Diarmuid " took him by one hand, and

Goll MacMornee by the other, and, pulling with all their might, tore him from the wall and the floor. But if they did, he left all the skin of his back, from his head to his heels, on the floor and the wall behind him. But when they were going home through the hills of Tralee, they found a sheep on the way, killed it, and clapped the skin on Conán. The sheepskin grew to his body; and he was so well and strong that they sheared him every year, and got wool enough from his back to make flannel and frieze for the Fenians of Erin ever after." This is a favourite incident, and recurs in the story of the laughing Gruagach. In most of the Fenian stories the fighting is brisk and incessant. It is quite a Donnybrook fair. Everybody kills everybody else, and then some toothless old woman comes along and rubs a magic salve on them, when, all in a minute, up they pop, and go at it again.

One of the quaintest conceits, and a pretty one withal, is that of Tir na n-Og, the Land of Youth, the life-giving region just beneath the ground, whence mysteriously spring the sturdy trees, the soft green grass, and the bright flowers. The journey thither is not long; sometimes the hero just pulls up a root and dives down through the hole into the blessed Tir na n-Og, — as primitive a

bit of folk-lore as one could wish to find! A
lovely country, of course, was that land of sprouting life, and some queer customs did they have
there. The mode of " running for office " was especially worthy of mention. Once in seven years
all the champions and best men " met at the front
of the palace, and ran to the top of a hill two
miles distant. On the top of that hill was a chair,
and the man that sat first in the chair was king of
Tir na n-Og for the next seven years." This method
enabled them to dispense with nominating conven-
tions and campaign lies, but not with intrigue and
sorcery, as we find in the droll story of Oisin (or
Ossian), which concludes the Fenian series.

The story of the Fisherman's Son and the Gruagach of Tricks is substantially the same with the
famous story of Farmer Weathersky, in the Norse
collection translated by Sir George Dasent. Gruagach (accented on the first syllable) means " the
hairy one," and, as Mr. Curtin cautiously observes,
" we are more likely to be justified in finding a
solar agent concealed in the person of the laughing
Gruagach or the Gruagach of Tricks than in many
of the sun myths put forth by some modern writers." He reminds one of Hermes and of Proteus,
and in the wonderful changes at the end of the
story we have, as in Farmer Weathersky, a vari-

ant of the catastrophe in the story of the Second
Royal Mendicant in the "Arabian Nights;" but the
Irishman gives us a touch of humour that is quite
his own. The Gruagach and his eleven artful
sons are chasing the fisherman's son through water
and air, and various forms of fish and bird are
assumed, until at length the fisherman's son, in the
shape of a swallow, hovers over the summerhouse
where the daughter of the king of Erin is sitting.
Weary with the chase, the swallow becomes a ring,
and falls into the girl's lap; it takes her fancy, and
she puts it on her finger. Then the twelve pur-
suers change from hawks into handsome men, and
entertain the king in his castle with music and
games, until he asks them what in the world he
can give them. All they want, says the old Gruagach, is the ring which he once lost, and which is
now on the princess's finger. Of course, says the
king, if his daughter has got the ring, she must give
it to its owner. But the ring, overhearing all this,
speaks to the princess, and tells her what to do.
She gets a gallon of wheat grains and three gallons
of the strongest *potheen* that was ever brewed in
Ireland, and she mixes them together in an open
barrel before the fire. Then her father calls her
and asks for the ring; and when she finds that her
protests are of no avail, and she must give it up,

she throws it into the fire. "That moment the eleven brothers made eleven pairs of tongs of themselves; their father, the old Gruagach, was the twelfth pair. The twelve jumped into the fire to know in what spark of it would they find the old fisherman's son; and they were a long time working and searching through the fire, when out flew a spark, and into the barrel. The twelve made themselves men, turned over the barrel, and spilled the wheat on the floor. Then in a twinkling they were twelve cocks strutting around. They fell to, and picked away at the wheat, to know which one would find the fisherman's son. Soon one dropped on one side, and a second on the opposite side, until all twelve were lying drunk from the wheat."

One seems to see the gleam in the corner of the eye and the pucker in the Gaelic visage of the old narrator. To be sure, it was the wheat. It could n't have been the mountain dew; it never is. Well, when things had come to this pass, the spark that was the fisherman's son just turned into a fox, and with one smart bite he took the head off the old Gruagach, and the eleven other boozy cocks he finished with eleven other bites. Then he made himself the handsomest man in Erin, and married the princess and succeeded to the crown.

There is a breezy freshness about these tales,

which will make the book a welcome addition to
young people's libraries.　It is safe to predict for
it an enviable success.　In the next edition there
ought to be an index, and we wish the author need
not feel it necessary to be so sparing with his own
notes and comments.　His brief Introduction is so
charming, from its weight of sense and beauty of
expression, that one would gladly hear more from
the author himself.　It is to be hoped that the
book lately published is the forerunner of many.

　August, 1890.

XII

GUESSING AT HALF AND MULTIPLYING BY TWO

" THE small philosopher is a great character in New England. His fundamental rule of logical procedure is to guess at the half and multiply by two. [Applause.] " [1] It is [in 1880] only two or three years since the philosopher from whom this text is quoted was himself a great character in New England, inasmuch as he could give a lecture once every week, in one of the largest halls of New England's principal city, and could entertain his audience of two or three thousand people with discussions of the most vast and abstruse themes of science and metaphysics. The success with which he entertained his audience is carefully chronicled for us in the volumes made up from the reports of his lectures, in which parenthetical notes

[1] Cook's *Boston Monday Lectures: Biology*, p. 51. After some hesitation I have decided to reprint this paper, because the " fundamental rule of procedure " here criticised is a favourite one with other controversialists than Mr. Cook, and it is one against which readers sometimes need to be put on their guard.

of " laughter," " applause," or " sensation " occur
as frequently as in ordinary newspaper reports of
stump speeches or humorous convivial harangues.
As a social phenomenon this career of the Rev.
Joseph Cook possesses considerable interest, —
enough, at any rate, to justify a brief inquiry as to
his " fundamental rule of procedure."

Among the wise and witty sayings of the ancients with which our children are puzzled and edi-
fied in the first dozen pages of the Greek Reader,
there is a caustic remark attributed to Phokion, on
the occasion of being very loudly applauded by the
populace. " Dear me," said the old statesman,
" can it be that I have been making a fool of myself? " So, when three thousand people are made
to laugh and clap their hands over statements about
the origin of species or the anatomy of the nervous
system, the first impulse of any scientific inquirer
of ordinary sagacity and experience is to ask in
what meretricious fashion these sober topics can
have been treated, in order to have produced such
a result. The inference may be cynical, but is
none the less likely to be sound. In the present
case, one does not need to read far in the published
reports of these lectures to cee that the fundamental
rule of procedure is something very different from
any of the rules by which truth is wooed and won

by scientific inquirers. Among Mill's comprehensive canons of logical method one might search in vain for a specimen of the method employed by Mr. Cook. Of the temper of mind, indeed, in which scientific inquiries are conducted, he has no more conception than Laura Bridgman could have of Pompeian red or a chord of the minor ninth. The process of holding one's judgment in suspense over a complicated problem, of patiently gathering and weighing the evidence on either side, of subjecting one's own first-formed hypotheses to repeated verification, of clearly comprehending and fairly stating opposing views, of setting forth one's conclusions at last, guardedly and with a distinct consciousness of the conditions under which they are tenable, — all this sort of thing is quite foreign to Mr. Cook's nature.

To him a scientific thesis is simply a statement over which it is possible to get up a fight. The gamecock is his totem ; to him the bones of the vertebrate subkingdom are only so many bones of contention, and the sponge is interesting chiefly as an emblem which is never, on any account, to be thrown up. He talks accordingly of scientific men lying in wait for Mr. Darwin, ready to pounce on him like a tiger on its prey ; he is very fond of exhibiting what he calls the " strategic point " of a

scientific book or theory ; and altogether his attitude is bellicose to a degree that is as unbecoming
in a preacher of the gospel as it is out of place in
a discussion of scientific questions.   His favourite
method of dealing with a scientific writer is to
quote from him all sorts of detached statements
and inferences, and, without the slightest regard
to the writer's general system of opinions or habits
of thought, to praise or vituperate the detached
statements according to some principle which it is
not always easy for the reader to discover, but
which has always doubtless some reference to their
supposed bearings upon the peculiar kind of orthodoxy of which Mr. Cook appears as the champion.
There are some writers whom he thinks it necessary always to scold or vilify, no matter what they
say.   If they happen to say something which ought
to be quite satisfactory to any reasonable person of
" orthodox " opinions, Mr. Cook either accuses them
of insincerity or represents them as making " concessions."

This last device, I am sorry to be obliged to add,
is not an uncommon one with theological controversialists, when their zeal runs away with them.
When a man makes a statement which expresses
his deepest convictions, there is no easier way of
seeming to knock away the platform on which he

stands than to quote his statement, and describe it
as something which he has reluctantly " conceded."
In dealing with the principal writers on evolution,
Mr. Cook is continually found resorting to this
cheap device. For example, when Professor Tyndall declares that " if a right-hand spiral movement of the particles of the brain could be shown
to occur in love, and a left-hand spiral movement
in hate, we should be as far off as ever from understanding the connection of this physical motion
with the spiritual manifestations," — when Professor Tyndall declares this, he simply asserts
what is a cardinal proposition with the group of
English philosophers to which he belongs. With
Professor Huxley, as well as with Mr. Spencer, it
is a fundamental proposition that psychical phenomena cannot possibly be interpreted in terms of
matter and motion, and this proposition they have
at various times set forth and defended. In the
chapter on Matter and Spirit, in my work on
" Cosmic Philosophy," I have fully expounded
this point, and have further illustrated it in
" The Unseen World." With the conclusions
there set forth the remark of Professor Tyndall
thoroughly agrees, and it does so because all these
expressions of opinion and all these arguments are
part and parcel of a coherent system of anti-mate-

rialistic thought adopted [1] by the English school
of evolutionists. Yet when Mr. Cook quotes
Professor Tyndall's remark, he does it in this wise :
" It is notorious that even Tyndall *concedes*," etc.,
etc.

By proceeding in this way, Mr. Cook finds it
easy to make out a formidable array of what he
calls " the concessions of evolutionists." He first
gives the audience a crude impression of some sort
of theory of evolution, such as no scientific thinker
ever dreamed of ; or, to speak more accurately,
he plays upon the crude impression already half
formed in the average mind of his audience, and
which he evidently shares himself. The *average*
notion of the doctrine of evolution, possessed in
common by an audience big enough to fill Tremont Temple, would no doubt seem to Darwin or
to Spencer something quite fearful and wonderful.
Playing with this sort of crude material, Mr. Cook
puts together a series of numbered propositions,
which remind one of those interminable auction
catalogues of Walt Whitman, which some of our
British cousins, more ardent than discriminating,
mistake for a truly American species of inspired
verse. In this long catena of statements, almost

---

[1] In spite of an occasional slip of the pen which may seem to
imply the contrary. See above, pp. 58–60.

everything is easily seen to disagree with the crude general impression to which the speaker appeals, and almost everything is accordingly set down as a "concession." And as the audience go out after the lecture, they doubtless ask one another, in amazed whispers, how it is that sensible men who make so many "concessions" can find it in their hearts to maintain the doctrine of evolution at all!

Sometimes Mr. Cook goes even farther than this, and, in the very act of quoting an author's declared opinions, expressly refuses to give him credit for them. Thus he has the hardihood to say: "Even Herbert Spencer, *who would be very glad to prove the opposite*,[1] says, in his Biology, 'The proximate chemical principles or chemical units — albumen, fibrine, gelatine, or the hypothetical proteine substance — cannot possess the property of forming the endlessly varied structures of animal forms.'" Mr. Cook here lays claim to a knowledge of his author's innermost thoughts and wishes which is quite remarkable. For a fit parallel one would have to cite the instance of the German who flogged his son for profanity, though the boy had not opened his mouth. "You dinks tamn," exclaimed the irate father, "and I vips you for dat!"

As there are some writers whom Mr. Cook

---

[1] The italicizing is, of course, mine, both here and below.

thinks it always necessary to vituperate, no matter
what they say, so there are others whom he finds it
convenient to quote, as foils to the former, and to
mention with praise on all occasions, though it is
difficult to assign the reasons for this preference,
except on the hypothesis that the lecturer has an
implicit faith in the simple and confiding nature
of his audience. Before giving these lectures Mr.
Cook had studied awhile in Germany, and his
citations of German writers show how far he deems
it safe to presume on New England's ignorance of
what the Fatherland thinks. It is nice to have
such a learned country as Germany at one's disposal to hurl at the heads of people whose "outlook in philosophy does not reach beyond the
Straits of Dover;" it saves a great deal of trouble-
some argument, and still more painful examination
of facts. This English opinion is all very well,
you know, but it comes from a philosopher "whose
star is just touching the western pines," and a German professor whom I am about to quote, whose
book I "hold in my hand," and "whose star is in
the ascendant," does not agree with it. All this
is extremely neat and convincing, apparently, to
the crowd in Tremont Temple. With all Germany at his disposal, however, it must be acknowledged that our lecturer makes a very sparing

use of his resources. He quotes Helmholtz and Wundt every now and then with warm approval, though wherein they should be found any more acceptable to the orthodox world than Tyndall and Spencer it is not easy to see, save that the ill repute of German free thinkers takes somewhat longer to get diffused in New England than the ill repute of English free thinkers.

Then, among these Germans who are to set the English-speaking world aright we have Delitzsch! To speak of Wundt and Delitzsch is as if one were to bracket together John Stuart Mill and Frederick Denison Maurice. And then comes the admirable Lotze, whom Mr. Cook is continually setting off as a foil to Herbert Spencer. On page 179 of the lectures on " Heredity " he enumerates, with emphasis, those opinions of Lotze which he deems of especial importance with regard to the relations between matter and mind, and then proceeds to deprecate the " thunder " which he presumes he has evoked " from all quarters of the Spencerian sky." But, considering that the propositions he quotes from Lotze express the very views of Herbert Spencer, only somewhat inadequately worded, it would seem that the lecturer's alarm cannot be very real, and the thunder in question is only a kind of comic-opera thunder manufactured behind the curtain for the benefit of the acquiescent audience. For example, the fourth proposition quoted with approval from Lotze reads thus : " Physical phenomena point to an underlying being to which they belong, but do not determine whether that being is material or immaterial." Now this is Spencerism, pure and simple, and it is a crucial proposition, too, pointing out the drift of the whole philosophy before which it is set up. The fact that Mr. Cook adopts such an opinion when stated by Lotze, but vituperates the same opinion when stated by Spencer, reveals to us, with a pungent though not wholly delicious flavour, the "true inwardness" of his fundamental method of procedure.

That method, it must be acknowledged with due regard to the *bon mot* of the old Greek statesman, is a method well adapted to conciliate the favour of an immense audience, — even in Boston. We are all descended from fighting ancestors, and many of us, who care little for the disinterested discussion of scientific theories, still like to see a man knocked down or impaled, provided the knocking down be done with a syllogistic club, or the impaling be restricted to such a hard substance as is afforded by the horns of a dilemma. It satisfies our combative instincts, without shocking our

physical sympathies or making any great demand
on our keener thinking powers, which most people
do most of all dislike to be called upon to exercise.
To this kind of feeling Mr. Cook's lectures appeal,
and the peculiar character of his success seems to
show that he knows well how to deal with it. In
a moment of winning frankness he exclaims: " Do
you suppose that I think that this audience can be
cheated? I do not know where in America there
is another weekly audience with as many brains in
it ; at least, I do not know where in New England
I should be so likely to be tripped up, if I were to
make an incorrect statement, as here." [1] After
this coaxing little dose, Mr. Cook proceeds to show
his respect for the learning of his audience in
some remarks on *bathybius*, which, as he conde-
scendingly explains, is a name derived from two
Greek words, meaning *deep* and *sea ! !* The profound knowledge of Greek thus exhibited is quite
equalled by his account of bathybius from the
zoölogical point of view. He begins by telling his
hearers that, in a paper published in the " Micro-
scopical Journal " in 1868, Professor Huxley " announced his belief that the gelatinous substance
found in the ooze of the beds of the deep seas is a
sheet of living matter extending around the globe."

[1] *Biology.* p. 67.

Furthermore, of " this amazingly strategic [!!] and
haughtily trumpeted substance . . . Huxley assumed that it was in the past, and would be in the
future, the progenitor of all the life on the planet."
Now it is not true that, in the paper referred to,
Huxley announces any such belief or makes any
such assumption as is here ascribed to him; but we
shall see, in a moment, that Mr. Cook's system of
quotation is peculiar in enabling him to extract
from the text of an author any meaning whatever
that may happen to suit his purposes.   This ingenious garbling enables the lecturer to come in with
telling effect at the close of his third lecture, and
earn an ignoble round of applause by holding up
the current number of the " American Journal of
Science and Arts "(which he would appear to have
picked up at a bookstall on his way to the lecture
room) and citing from it, as the fifty-first and clos-
ing " concession " of evolutionists, " that bathybius
has been discovered in 1875, by the ship Challenger,
to be — hear, O heavens ! and give ear, O earth ! —
sulphate of lime ; and that when dissolved it crys-
tallizes as gypsum.   [Applause.] "   This is what
Mr. Cook calls striking, with the " latest scientific
intelligence," at the " bottom stem " of the great
tree of evolution.   The " latest scientific intelligence," with him, means the last book or article

which he has glanced over without comprehending
its import, but from which he has contrived to
glean some statement calculated to edify his audience and scatter the hosts of Midian. In point of
fact, the identification of bathybius with sulphate
of lime was set down by Sir Wyville Thomson only
as a suspicion, to which Huxley, like a true man of
science, at once accorded all possible weight, while
leaving the question open for further discussion.
Only a mountebank, dealing with an audience upon
whose ignorance of the subject he might safely
rely, could pretend to suppose that the fate of the
doctrine of evolution was in any way involved in
the question as to the organic nature of bathybius.
The amazing strategy was all Mr. Cook's own,
and the haughty trumpeting appears to have been
chiefly done with his own very brazen instrument.

I said a moment ago that Mr. Cook's system of
quotation is peculiar. The following instance is
so good that it will bear citing at some length.
According to Mr. Cook, Professor Huxley says, in
his article on Biology in the ninth edition of the
" Encyclopædia Britannica : " " *Throughout almost the whole series of living beings, we find
agamogenesis, or not-sexual generation.*" After
a pause, Mr. Cook proceeded in a lower voice :
" When the topic of the origin of the life of our

Lord on the earth is approached from the point of view of the microscope, some men, who know not what the holy of holies in physical and religious science is, say that we have no example of the origin of life without two parents." He went on to cite the familiar instances of parthenogenesis in bees and silk moths, and then proceeded as follows: " Take up your Mivart, your Lyell, your Owen, and you will read [where?] this same important fact which Huxley here asserts, when he says that the law that perfect individuals may be virginally born extends to the higher forms of life. I am in the presence of Almighty God ; and yet, when a great soul like that tender spirit of our sainted Lincoln, in his early days, with little knowledge but with great thoughtfulness, was troubled by this difficulty, and almost thrown into infidelity by not knowing that the law that there must be two parents is not universal, I am willing to allude, even in such a presence as this, to the latest science concerning miraculous conception.   [Sensation.] "

The vulgarity of this rhetoric is as glaring as its absurdity. All that concerns me now, however, is to point out the Brobdignagian dimensions of the misstatement of facts. Let us look back for a moment at the italicized quotation from Huxley, upon which Mr. Cook builds up the wondrous assertion " that the law that perfect individuals may be virginally born extends to the higher forms of life." Then let us turn to Huxley's article and see what he really does say.

Treating of the whole subject of agamogenesis in the widest possible way by including it under the more general process of cell-multiplication, Huxley says : " Common as the process is in plants and in the lower animals, it becomes rare among the higher animals. In these, the reproduction of the whole organism from a part, in the way indicated above, ceases. At most we find that the cells at the end of an amputated portion of the organism are capable of reproducing the lost part, and in the very highest animals even this power vanishes in the adult. . . . *Throughout almost the whole series of living beings, however, we find concurrently with the process of agamogenesis, or asexual generation,* another method of generation, in which the development of the germ into an organism resembling the parent depends on an influence exerted by living matter different from the germ. This is *gamogenesis,* or sexual generation."[1]

Comparing the italicized passage here with Mr. Cook's italicized quotation, we see vividly illustrated the fundamental method of procedure

[1] *Encyclopædia Britannica,* ninth edition, " Biology," p. 686.

by which the " Monday Lectureship " jumps from a statement about the reproduction of a lobster's claw to the inference that a man may be born without a father. It reminds one of that worthy clergyman who introduced a scathing sermon on a new-fangled variety of ladies' headdress by the appropriate text, " Top-knot come down ! " On being reminded by one of his deacons that the full verse seemed to read, " Let him that is upon the housetop not come down," the pastor boldly justified his abridgment on the ground that any particular collocation of words in Scripture is as authoritative as any other, since all parts of the Bible are equally inspired. Perhaps there are some who would justify Mr. Cook's peculiar principle of abridgment on the familiar ground that the end sanctifies the means, and that if a statement seems helpful to " the truth " in general, it is no matter whether the statement itself is true or not.

Enough of this. If we were to go through with these volumes in detail, we should find little else but misrepresentations of facts, misconceptions of principles, and floods of tawdry rhetoric, of which the specimens here quoted are quite sufficient to illustrate the lecturer's " fundamental method of procedure." If I have treated him somewhat lightly, it is because there is nothing in his matter or in his manner that would justify, or even excuse, a more serious style of treatment. The only aspect of his career which affords matter for grave reflection is the ease with which he succeeded for a moment in imposing on the credulity and in appealing to the prejudices of his public. The eagerness with which the orthodox world hailed the appearance of this new champion could not but remind one, with sad emphasis, of Oxenstjern's famous remark : "Quam parva sapientia mundus regitur!" It is comforting to remember that one of the world's greatest naturalists, Asa Gray, — whose orthodoxy is as unimpeachable as his science, — very promptly declared in print that such championship is something of which orthodoxy has no reason to feel proud.

December, 1880.

XIII

FORTY YEARS OF BACON-SHAKESPEARE FOLLY [1]

SOME time ago, while I happened to be looking over a wheelbarrow-load of rubbish written to prove that such plays as " King Lear " and " The Merry Wives of Windsor " emanated from one of the least poetical and least humorous minds of modern times, I was reminded of a story which I heard when a boy. I forget whether it was some whimsical man of letters like Charles Lamb, or some such professional wag as Theodore Hook, who took it into his head one day to stand still on a London street, with face turned upward, gazing into the sky. Thereupon the next person who came that way forthwith stopped and did likewise, and then the next, and the next, until the road was blocked by a dense crowd of men and women, all standing as if rooted in the ground, and with solemn skyward stare. The enchantment was at last broken

---

[1] This article was published in the fortieth-anniversary number of *The Atlantic Monthly*, November, 1897.

when some one asked what they were looking at,
and nobody could tell.   It was simply an instance
of a certain remnant of primitive gregariousness of
action on the part of human beings, which exhibits
itself from time to time in sundry queer fashions
and fads.

So when Miss Delia Bacon, in the year which
saw the beginning of " The Atlantic Monthly,"
published a book purporting to unfold the " philosophy " of Shakespeare's dramas, it was not long
before other persons began staring intently into
the silliest mare's nest ever devised by human dul-
ness ; and the fruits of so much staring have appeared in divers eccentric volumes, of which more
specific mention will presently be made.   Neither
in number nor in quality are they such as to indicate that the Bacon - Shakespeare folly has yet
become fashionable, and we shall presently observe
in it marked suicidal tendencies which are likely
to prevent its ever becoming so ; but there are
enough of such volumes to illustrate the point of
my anecdote.

Another fad, once really fashionable, and in defence of which some plausible arguments could be
urged, was the Wolfian theory of the Homeric
poems, which dazzled so many of our grandparents.
It is worth our while to mention it here by way of

prelude.   The theory that the Iliad and Odyssey
are mere aggregations of popular ballads, collected
and arranged in the time of Pisistratus, was perhaps originally suggested by the philosopher Vico,
but first attracted general attention in 1795, when
set forth by Friedrich August Wolf, one of the
most learned and brilliant of modern scholars.
Thus eminently respectable in its parentage and
quite reasonable on the surface, this ballad theory
came to be widely fashionable; forty years ago it
was accepted by many able scholars, though usually with large modifications.

The Wolfians urged that we know absolutely
nothing about the man Homer, not even when or
where he lived.   His existence is merely matter of
tradition, or of inference from the existence of the
poems.   But as the poems know nothing of Do-
rians in Peloponnesus, their date can hardly be
so late as 1100 B. C.   What happened, then, when
" an edition of Homer " was made at Athens,
about 530 B. C., by Pisistratus, or under his orders?
Did the editor simply edit two great poems already
six centuries old, or did he make up two poems
by piecing together a miscellaneous lot of ancient
ballads?   Wolf maintained the latter alternative,
chiefly because of the alleged impossibility of com-
posing and preserving such long poems in the

alleged absence of the art of writing. Having thus made a plausible start, the Wolfians proceeded to pick the poems to pieces, and to prove by "internal evidence" that there was nothing like "unity of design" in them, etc.; and so it went on, till poor old Homer was relegated to the world of myth. As a schoolboy I used to hear the belief in the existence of such a poet derided as "uncritical" and "unscholarly."

In spite of these terrifying epithets, the ballad theory never made any impression upon me; for it seemed to ignore the most conspicuous and vital fact about the poems, namely, the *style*, the noble, rapid, simple, vivid, supremely poetical style, — a style as individual and unapproachable as that of Dante or Keats. For an excellent characterization of it, read Matthew Arnold's charming essays "On Translating Homer." The style is the man, and to suppose that this Homeric style ever came from a democratic multitude of minds, or from anything save one of those supremely endowed individual natures such as get born once or twice in a millennium, is simply to suppose a psychological impossibility. I remember once talking about this with George Eliot, who had lately been reading Frederick Paley's ingenious restatement of the ballad theory, and was captivated by its ingenuity. I told her I did not wonder that old dry-asdust philologists should hold such views, but I was indeed surprised to find such a literary artist as herself ignoring the impassable gulf between Homer's language and that which any ballad theory necessarily implies. She had no answer for this except to say that she should have supposed an evolutionist like me would prefer to regard the Homeric poems as gradually evolved rather than suddenly created! A retort so clever and amiable most surely entitled her to the woman's privilege of the last word.

The Wolfian theory may now be regarded as a thing of the past; it has had its day and been flung aside. If Wolf himself were living, he would be the first to laugh at it. Its original prop has been knocked away, since it has become pretty clear that the art of writing was practised about the shores of the Ægean Sea long before 1100 B. C. Even Wolf would now admit that it might have been a real letter that Bellerophon carried to the father of Anteia.[1] All attempts to show a lack of unity in the design of the Iliad and the Odyssey have failed irretrievably, and the discussion has served only to make more and more unmistakable the work of the mighty master. The

[1] Iliad, vi. 168.

ballad theory is dead and buried, and he who would read its obituary may find keen pleasure, as well as many a wholesome lesson in sound criticism, in the sensible and brilliant book by Andrew Lang on " Homer and the Epic."

The Bacon-Shakespeare folly has never been set forth by scholars of commanding authority, like Wolf and Lachmann, or Niese and Wilamowitz Moellendorff. Among Delia Bacon's followers not one can by any permissible laxity of speech be termed a scholar, and their theory has found acceptance with very few persons. Nevertheless, it illustrates as well as the Wolfian theory the way in which such notions grow. It starts from a false premise, hazily conceived, and it subsists upon arguments in which trivial facts are assigned higher value than facts of vital importance. Mr. Lang's remark upon certain learned Homeric commentators, " that they pore over the hyssop on the wall, but are blind to the cedar of Lebanon," applies with tenfold force to the Bacon-Shakespeare sciolists. In them we always miss the just sense of proportion which is one of the abiding marks of sanity. The unfortunate lady who first brought their theory into public notoriety in 1857 was then sinking under the cerebral disease of which she died two years later, and

her imitators have been chiefly weak minds of the
sort that thrive upon paradox, closely akin to the
circle-squarers and inventors of perpetual motion.
Underlying all the absurdities, however, there is
something that deserves attention. Like many
other morbid phenomena, the Bacon-Shakespeare
folly has its natural history which is instructive.
The vagaries of Delia Bacon and her followers
originated in a group of conditions which admit of
being specified and described, and which the historian of nineteenth-century literature will need to
notice. In order to understand the natural history
of the affair, it is necessary to examine the Delia
Bacon theory at greater length than it would otherwise deserve. Let us see how it is constructed.

It starts with a syllogism, of which the major
premise is that the dramas ascribed to Shakespeare
during his lifetime, and ever since believed to be
his, abound in evidences of extraordinary book-
learning. The minor premise is that William
Shakespeare of Stratford-on-Avon could not have
acquired or possessed so much book-learning. The
conclusion is that he could not have written those
plays.

The question then arises, Which of Shakespeare's contemporaries had enough book-lore to
have written them? No doubt Francis Bacon had

enough. The conclusion does not follow, however, that he wrote the plays; for there were other contemporaries with learning enough and to spare, as for example George Chapman and Ben Jonson. These two men, to judge from their acknowledged works, were great poets, whereas in Bacon's fifteen volumes there is not a paragraph which betrays poetical genius. Why not, then, ascribe the Shakespeare dramas to Chapman or Jonson? Here the Baconizers endeavour to support their assumption by calling attention to similarities in thought and phrase between Francis Bacon and the writer of the dramas. Up to this point their argument consists of deductions from assumed premises; here they adduce inductive evidence, such as it is. We shall see specimens of it by and by. At present we are concerned with the initial syllogism.

And first, as to the major premise, it must be met with a flat denial. The Shakespeare plays do not abound with evidences of scholarship or learning of the sort that is gathered from profound and accurate study of books. It is precisely in this respect that they are conspicuously different from many of the plays contemporary with them, and from other masterpieces of English literature. Such plays as Jonson's "Sejanus" and "Catiline" are the work of a scholar deeply indoctrinated with

the views and mental habits of classic antiquity;
he has **soaked** himself **in** the **style of** Lucan and
Seneca, **until** their mental peculiarities have be-
come like a second nature to him, and are unconsciously betrayed alike in the general handling of
his **story and** in little turns of expression. Or take
Milton's " Lycidas : " no one but a man saturated
in every fibre with Theocritus and Virgil could
have written such a poem. An extremely foreign
and artificial literary form has been so completely
mastered and assimilated by **Milton** that he uses it
with as much ease as Theocritus himself, and has
produced a work that even the master of idyls had
scarcely equalled. After the terrific invective
against the clergy and the beautiful invocation to
the flowers, followed by the triumphant hallelujah
of Christian faith, observe the sudden reversion to
pagan sentiment where Lycidas is addressed as the
genius of the shore. Only profound scholarship
could have written this wonderful poem, — could
have brought forth the Christian thought as if
spontaneously through the medium of the pagan
form. Now there is nothing of this sort in Shakespeare. He uses classical materials, or anything
else under the sun that suits his purpose. He
takes a chronicle from Holinshed, a biography from
North's translation of Plutarch, a legend from

Saxo Grammaticus through Belleforest's French version, a novel of Boccaccio, a miracle play, — whatever strikes his fancy; he chops up his materials and weaves them into a story without much regard to classical models; defying rules of order and unity, and not always heeding probability, but never forgetful of his abiding purpose, to create live men and women. These people may have Greek and Latin names, and their scene of action may be Rome or Mitylene, decorated with scraps of classical knowledge such as a bright man might pick up in miscellaneous reading; but all this is the superficial setting, the mere frame to the picture. The living canvas is human nature as Shakespeare saw it in London and depicted with supreme poetic faculty. Among the new books within his reach was Chapman's magnificent translation of the Iliad, which at a later day inspired Keats to such a noble outburst of encomium; and in "Troilus and Cressida" we have the Greek and Trojan heroes set before us with an incisive reality not surpassed by Homer himself. This play shows how keenly Shakespeare appreciated Homer, how delicately and exquisitely he could supplement the picture; but there is nothing in its five acts that shows him clothed in the garment of ancient thought as Milton wore it. Shakespeare's freedom from such lore is a great advantage to him; in "Troilus and Cressida" there is a freedom of treatment hardly possible to a professional scholar. It is because of this freedom that Shakespeare reaches a far wider public of readers and listeners than Milton or Dante, whose vast learning makes them in many places "caviare to the general." Book-lore is a great source of power, but one may easily be hampered by it. What we forever love in Homer is the freshness that comes with lack of it, and in this sort of freshness Shakespeare agrees with Homer far more than with the learned poets.

It is not for a moment to be denied that Shakespeare's plays exhibit a remarkable wealth of varied knowledge. The writer was one of the keenest observers that ever lived. In the woodland or on the farm, in the printing shop or the alehouse, or up and down the street, not the smallest detail escaped him. Microscopic accuracy, curious interest in all things, unlimited power of assimilating knowledge, are everywhere shown in the plays. These are some of the marks of what we call *genius*, — something that we are far from comprehending, but which experience has shown that books and universities cannot impart. All the colleges on earth could not by combined effort make the kind of man we call a genius, but such a man may at any moment be born into the world, and it is as likely to be in a peasant's cottage as anywhere.

There is nothing in which men differ more widely than in the capacity for imbibing and assimilating knowledge. The capacity is often exercised unconsciously. When my eldest son, at the age of six, was taught to read in the course of a few weeks of daily instruction, it was suddenly discovered that his four-year-old brother also could read. Nobody could tell how it happened. Of course the younger boy must have taken keen notice of what the elder one was doing, but the process went on without attracting attention until the result appeared.

This capacity for unconscious learning is not at all uncommon. It is possessed to some extent by everybody; but a very high degree of it is one of the marks of genius. I remember one evening, many years ago, hearing Herbert Spencer in a friendly discussion regarding certain functions of the cerebellum. Abstruse points of comparative anatomy and questions of pathology were involved. Spencer's three antagonists were not violently opposed to him, but were in various degrees unready to adopt his views. The three were: Huxley, one of the greatest of comparative anatomists; Hughlings Jackson, a very eminent authority on the pathology of the nervous system; and George

Henry Lewes, who, although more of an amateur
in such matters, had nevertheless devoted years of
study to neural physiology, and was thoroughly
familiar with the history of the subject. Spencer
more than held his ground against the others. He
met fact with fact, brought up points in anatomy
the significance of which Huxley confessed that he
had overlooked, and had more experiments and
clinical cases at his tongue's end than Jackson
could muster. It was quite evident that he knew
all they knew on the subject, and more besides.
Yet Spencer had never been through a course of
"regular training" in the studies concerned; nor
had he ever studied at a university, or even at a
high school. Where did he learn the wonderful
mass of facts which he poured forth that evening?
Whence came his tremendous grasp upon the principles involved? Probably he could not have told
you. A few days afterward I happened to be talking with Spencer about history, a subject of which
he modestly said he knew but little. I told him I
had often been struck with the aptness of the historic illustrations cited in many chapters of his
"Social Statics," written when he was twenty-nine
years old. The references were not only always
accurate, but they showed an intelligence and
soundness of judgment unattainable, one would

think, save by close familiarity with history.
Spencer assured me that he had never read extensively in history. Whence, then, this wealth
of knowledge, — not smattering, not sciolism, but
solid, well-digested knowledge? Really, he did not
know, except that when his interest was aroused in
any subject he was keenly alive to all facts bearing
upon it, and seemed to find them whichever way
he turned. When I mentioned this to Lewes,
while recalling the discussion on the cerebellum,
he exclaimed: " Oh, you can't account for it! It's
his genius. Spencer has greater instinctive power
of observation and assimilation than any man since
Shakespeare, and he is like Shakespeare for hit-
ting the bull's-eye every time he fires. As for
Darwin and Huxley, we can follow their intellectual processes, but Spencer is above and beyond all;
he is inspired!"

Those were Lewes's exact words, and they made
a deep impression upon me. The comparison with
Shakespeare struck me as a happy one, and I can
understand both Spencer and Shakespeare the better for it. Concerning Spencer one circumstance
may be observed. Since his early manhood he has
lived in London, and has had for his daily associates men of vast attainments in every department
of science. He has thus had rare opportunities for

absorbing an immense fund of knowledge unconsciously.

It is evident that the author of Shakespeare's plays possessed an extraordinary " instinctive power of observation and assimilation." There was nothing strange in such a genius growing up in a small Warwickshire town. The difficulty is one which the Delia-Baconians have created for themselves. As it is their chief stock in trade, they magnify it in every way they can think of. Shakespeare's parents, they say, were illiterate, and he did not know how to spell his own name. It appears as Shagspere, Shaxpur, Shaxberd, Chacsper, and so on through some thirty forms, several of which William Shakespeare himself used indifferently. The implication is that such a man must have been shockingly ignorant. The real ignorance, however, is on the part of those who use such an argument. Apparently, they do not know that in Shakespeare's time such laxity in spelling was common in all ranks of society and in all grades of culture. The name of Elizabeth's great Lord Treasurer, Cecil, and his title, Burghley, were both spelled in half a dozen ways. The name of Raleigh occurs in more than forty different forms, and Sir Walter, one of the most accomplished men of his time, wrote it Rauley, Rawleyghe, Ralegh, and in yet other ways. The talk of the Baconizers on this point is simply ludicrous.

Equally silly is their talk about the dirty streets of Stratford. They seem to have just discovered that Elizabeth's England was a badly drained country, with heaps of garbage in the streets. Shakespeare's father, they tell us, was a butcher, and evidently from a butcher's son, living in an ill-swept town, and careless about the spelling of his name, not much in the way of intellectual achievement was to be expected! In point of fact, Shakespeare's parents belonged to the middle class. His father owned several houses in Stratford and two or three farms in the neighbourhood. As a farmer in those days, he would naturally have cattle slaughtered on his premises and would sell wool off the backs of his own flocks, whence the later tradition of his having been butcher and wool dealer. That his social position was good is shown by the facts that he was chief alderman and high bailiff of Stratford, and justice of the peace, and was styled " Master John Shakespeare," or (as we should say) " Mr. ; " whereas, had he been one of the common folk, his style had been " Goodman Shakespeare." A visit to his home in Henley Street, and to Anne Hathaway's cottage at Shottery, shows that the two families were in eminently respectable circumstances. The son of the high bailiff would see the best people in the neighbourhood. There was in the town a remarkably good free grammar school, where he might have learned the "small Latin and less Greek" which his friend Ben Jonson assures us he possessed. This expression, by the way, is usually misunderstood, because people do not pause to consider it. Coming from Ben Jonson, I should say that "small Latin and less Greek" might fairly describe the amount of those languages ordinarily possessed by a member of the graduating class at Harvard in good standing. It can hardly imply less than the ability to read Terence at sight, and perhaps Euripides less fluently. The author of the plays, with his unerring accuracy of observation, knows Latin enough at least to use the Latin part of English most skilfully; at the same time, when he has occasion to use Greek authors, such as Homer or Plutarch, he usually prefers an English translation. At all events, Jonson's remark informs us that the man whom he addresses as "sweet swan of Avon" knew *some* Latin and *some* Greek, — a conclusion which is so distasteful to one of our Baconizers, Mr. Edwin Reed, that he will not admit it. Rather than do so, he has the assurance to ask us to believe that by the epithet "sweet swan of Avon" Jonson really meant Francis Bacon! Dear me, Mr. Reed, do you really mean it? And how about the editor of Beaumont and Fletcher in 1647, when, in his dedication to Shakespeare's friend the Earl of Pembroke, he speaks of "Sweet Swan of Avon Shakespear"? Was he too a participator in the little scheme for fooling posterity? Or was he one of those who were fooled?

Whether Shakespeare had other chances for book-lore than those which the grammar school afforded, whether there was any interesting parson at hand, as often in small towns, to guide and stimulate his unfolding thoughts, — upon such points we have no information. But there were things to be learned in the country town quite outside of books and pedagogues. There, while the poet listened to the "strain of strutting chanticleer," and watched the "sun-burn'd sicklemen, of August weary," putting on their rye-straw hats and making holiday with rustic nymphs, he could rejoice in

> " Earth's increase, foison plenty,
> Barns and garners never empty;
> Vines with clust'ring bunches growing;
> Plants with goodly burthen bowing; "

there he could see the "unbacked colts" prick their ears, advance their eyelids, lift up their noses,

as if they smelt music; there he knew, doubtless,
many a bank where the wild thyme grew and on
which the moonlight sweetly slept; there he watched
the coming of "violets dim," "pale primroses,"
flower-de-luce, carnations, with "rosemary and rue"
to keep their "savour all the winter long,"

> "When icicles hang by the wall,
>   And Dick the shepherd blows his nail,
> And Tom bears logs into the hall,
>   And milk comes frozen home in pail."

Such lore as this no books or college could impart.

It was this that Milton had in mind when he
introduced Shakespeare and Ben Jonson into his
poem "L'Allegro." Milton was in his thirtieth
year when Jonson, poet laureate, was laid to rest
in Westminster Abbey; he was only a boy of
eight years when Shakespeare died, but the beautiful sonnet written fourteen years later shows how
lovingly he studied his works: —

> "What needs my Shakespeare, for his honoured bones," etc.

The poem "L'Allegro" and its fellow "Il Pense-
roso" describe the delights of Milton's life at his
father's country house near Windsor Castle. He
used often to ride into London to hear music or
pass an evening at the theatre, as in the following
lines: —

" Then to the well-trod stage anon,
   If Jonson's learned sock be on,
   Or sweetest Shakespeare, fancy's child,
   Warble his **native** woodnotes wild."

This accurate and happy contrast exasperates the Baconizers, for it spoils their stock in trade, and accordingly they try their best to assure us that Milton did not know what he was writing about. They asseverate with vehemence that in all the seven-and-thirty plays there is no such thing as a native woodnote wild.

But before leaving the contrast we may pause for a moment to ask, Where did Ben Jonson get his learning? He was, as he himself tells us, "poorly brought up" by his stepfather, a bricklayer. He went to Westminster School, where he was taught by Camden, and he may have spent a short time at Cambridge, though this is doubtful. His schooling was nipped in the bud, for he had to go home and lay brick; and when he found such an existence insupportable he went into the army and fought in the Netherlands. At about the age of twenty we find him back in London, and there lose sight of him for five years, when all at once his great comedy "Every Man in his Humour" is performed, and makes him famous. Now, in such a life, when did Jonson get the time for his immense reading and his finished classical scholar-

ship?   Reasoning after the manner of the Delia-
Baconians, we may safely say that he could not
possibly have accumulated the learning which is
shown in his plays : therefore he could not have
written those plays ; therefore Lord Bacon must
have written them !   There are daring soarers in
the empyrean who do not shrink from this conclusion ; a doctor in Michigan, named Owen, has published a pamphlet to prove, among other things,
that Bacon was the author of the plays which were
performed and printed as Jonson's.

To return to Shakespeare.   Somewhere about
1585, when he was one-and-twenty, he went to
London, leaving his wife and three young children
at Stratford.   His father had lost money, and the
fortunes of the family were at the lowest ebb.   In
London we lose sight of Shakespeare for a while,
just as we lose sight of Jonson, until literary works
appear.   The work first published is " Venus and
Adonis," one of the most exquisite pieces of dic-
tion in the English language.   It was dedicated to
Henry, Earl of Southampton, by William Shakespeare, whose authorship of the poem is asserted as
distinctly as the title-page of " David Copperfield "
proclaims that novel to be by Charles Dickens, yet
some precious critics assure us that Shakespeare
" could not " have written the poem, and never

knew the Earl of Southampton. Some years ago, Mr. Appleton Morgan, who does not wish to be regarded as a Baconizer, published an essay on the Warwickshire dialect, in which he maintained that since no traces of that kind of speech occur in " Venus and Adonis," therefore it could not have been written by a young man fresh from a small Warwickshire town. This is a specimen of the loose kind of criticism which prepares soil for Delia-Baconian weeds to grow in. The poem was published in 1593, seven or eight years after Shakespeare's coming to London ; and we are asked to believe that the world's greatest genius, one of the most consummate masters of speech that ever lived, could tarry seven years in the city without learning how to write what Hosea Biglow calls " citified English " ! One can only exclaim with Gloster, " O monstrous fault, to harbour such a thought ! "

In those years Shakespeare surely learned much else. It seems clear that he had a good reading acquaintance with French and Italian, though he often uses translations, as for instance Florio's version of Montaigne. In estimating what Shakespeare " must have " known or " could not have " known, one needs to use more caution than some of our critics display. For example, in " The

Winter's Tale" the statue of Hermione is called
" a piece . . . now newly performed by that rare
Italian master, Julio Romano." Now, since Romano is known as a great painter, but not as a
sculptor, this has been cited as a blunder on
Shakespeare's part. It appears, however, that the
first edition of Vasari's "Lives of the Painters,"
published in 1550 and never translated from its
original Italian, informs us that Romano did work
in sculpture. In Vasari's second edition, published in 1568 and translated into several languages, this information is not given. From these
facts, the erudite German critic Dr. Karl Elze,
who is not a bit of a Delia-Baconian, but only an
occasional sufferer from *vesania commentatorum*,
introduces us to a solemn dilemma : either the
author of " The Winter's Tale" must have consulted the first edition of Vasari in the original Italian, or else he must have travelled in Italy and
gazed upon statues by Romano. Ah ! prithee not
so fast, worthy doctor ; be not so lavish with these
" musts." It is, I think, improbable that Shakespeare ever saw Italy except with the eyes of his
imperial fancy. On the other hand, there are
many indications that he could read Italian, but
among them we cannot attach much importance to
this one. Why should he not have learned from

hearsay that Romano had made statues? In the name of common sense, are there no sources of knowledge save books? Or, since it was no unusual thing for Italian painters in the sixteenth century to excel in sculpture and architecture, why should not Shakespeare have assumed without verification that it was so in Romano's case? It was a tolerably safe assumption to make, especially in an age utterly careless of historical accuracy, and in a comedy which provides Bohemia with a seacoast, and mixes up times and customs with as scant heed of probability as a fairy tale.

In arguing about what Shakespeare " must have " or " could not have " known, we must not forget that at no time or place since history began has human thought fermented more briskly than in London while he was living there. The age of Drake and Raleigh was an age of efflorescence in dramatic poetry, such as had not been seen in the twenty centuries since Euripides died. Among Shakespeare's fellow craftsmen were writers of such great and varied endowments as Chapman, Marlowe, Greene, Nash, Peele, Marston, Dekker, Webster, and Cyril Tourneur. During his earlier years in London, Richard Hooker was master of the Middle Temple, and there a little later Ford and Beaumont were studying. The erudite Cam-

den was master of Westminster School; among
the lights of the age for legal learning were Edward Coke and Francis Bacon; at the same time,
one might have met in London the learned archi-
tect Inigo Jones and the learned poet John Donne,
both of them excellent classical scholars; there one
would have found the divine poet Edmund Spenser, just come over from Ireland to see to the publication of his "Faerie Queene;" not long afterward came John Fletcher from Cambridge, and the
acute philosopher Edward Herbert from Oxford.
and one and all might listen to the incomparable
table-talk of that giant of scholarship, John Selden. The delights of the Mermaid Tavern. where
these rare wits were wont to assemble, still live in
tradition. As Keats says: —

> "Souls of poets dead and gone,
> What Elysium have ye known,
> Happy field or mossy cavern,
> Choicer than the Mermaid Tavern?"

It has always been believed that this place was
one of Shakespeare's favourite haunts. By common consent of scholars, it has been accepted as
the scene of those contests of wit between Shakespeare and Jonson of which Fuller tells us when
he compares Jonson to a Spanish galleon, built
high with learning, but slow in movement, while

he likens Shakespeare to an English cruiser, less heavily weighted, but apt for victory because of its nimbleness, — the same kind of contrast, by the way, as that which occurred to Milton.

But our Baconizing friends will not allow that Shakespeare ever went to the Mermaid, or knew the people who met there ; at least, none but a few fellow dramatists. We have no documentary proof that he ever met with Raleigh, or Bacon, or Selden. Let us observe that, while these sapient critics are in some cases ready to welcome the slightest circumstantial evidence, there are others in which they will accept nothing short of absolute demonstration. Did Shakespeare ever see a may-pole ? The word occurs just once in his plays, namely, in the " Midsummer Night's Dream," where little Hermia, quarrelling with tall Helena, calls her a " painted maypole ;" but that proves nothing. I am not aware that there is any absolute docu-mentary proof that Shakespeare ever set eyes on a maypole. It is nevertheless certain that in Eng-land, at that time, no boy could grow to manhood without seeing many a maypole. Common sense has some rights which we are bound to respect.

Now, Shakespeare's London was a small city of from 150,000 to 200,000 souls, or about the size of Providence or Minneapolis at the present time.

In cities of such size, everybody of the slightest eminence is known all over town, and such persons are sure to be more or less acquainted with one an- other ; it is a very rare exception when it is not so. Before his thirtieth year, Shakespeare was well known in London as an actor, a writer of plays, and the manager of a prominent theatre. It was in that year that Spenser, in his " Colin Clout 's Come Home Again," alluding to Shakespeare under the name of Aëtion, or " eagle-like," paid him this compliment : —

> " And there, though last, not least, is Aëtion ;
> A gentler shepherd may nowhere be found ;
> Whose muse, full of high thought's invention,
> Doth, like himself, heroically sound."

Four years after this, in 1598, Francis Meres pub- lished his book entitled " Palladis Tamia," a very interesting contribution to literary history. The author, who had been an instructor in rhetoric in the University of Oxford, was then living in Lon- don, near the Globe Theatre. In this book Meres tells his readers that " the sweet witty soul of Ovid lives in mellifluous and honey-tongued Shakespeare ; witness his ' Venus and Adonis,' his ' Lucrece,' his sugared sonnets among his private friends, etc. . . . As Plautus and Seneca are accounted the best for comedy and tragedy among the Latins, so Shakespeare among the English is the most excellent in both kinds for the stage : for comedy, witness his 'Gentlemen of Verona,' his 'Errors,' his 'Love's Labour's Lost,' his 'Love's Labour's Wonne,' [1] his 'Midsummer Night's Dream,' and his 'Merchant of Venice;' for tragedy, his 'Richard II.,' 'Richard III.,' 'Henry IV.,' 'King John,' 'Titus Andronicus,' and his 'Romeo and Juliet.' As Epius Stolo said that the Muses would speak with Plautus's tongue if they would speak Latin, so I say that the Muses would speak with Shakespeare's fine filed phrase if they would speak English." In other passages Meres mentions Shakespeare's lyrical quality, for which he likens him to Pindar and Catullus ; and the glory of his style, for which he places him along with Virgil and Homer. It thus appears that, at the age of thirty-four, this poet from Stratford was already ranked by critical scholars by the side of the greatest names of antiquity. Let me add that the popularity of his plays was making him a somewhat wealthy man, so that he had relieved his father from pecuniary troubles, and had just bought for himself the Great House at Stratford where the last years of his life were spent. His income seems already to have been equivalent to $10,000 a year in our modern money. His posi-

---

[1] The comedy afterward developed into *All's Well that Ends Well.*

tion had come to be such that he could extend
patronage to others. It was in 1598 that through
his influence Ben Jonson obtained, after many rebuffs, his first hearing before a London audience,
when "Every Man in his Humour" was brought
out at Blackfriars Theatre, with Shakespeare acting one of the parts.

To suppose that such a man as this, in a town
the size of Minneapolis, connected with a principal theatre, writer of the most popular plays of the
day, a poet whom men were already coupling with
Homer and Pindar, — to suppose that such a man
was not known to all the educated people in the
town is simply absurd. There were probably very
few men, women, or children in London, between
1595 and 1610, who did not know who Shakespeare was when he passed them in the street ; and
as for such wits as drank ale and sack at the Mermaid, as for Raleigh and Bacon and Selden and
the rest, to suppose that Shakespeare did not know
them well — nay, to suppose that he was not the
leading spirit and brightest wit of those ambrosial
nights — is about as sensible as to suppose that he
never saw a maypole.

The facts thus far contemplated point to one
conclusion. The son of a well-to-do magistrate in
a small country town is born with a genius which

the world has never seen surpassed. Coming to London at the age of twenty-one, he achieves such swift success that within thirteen years he is recognized as one of the chief glories of English literature. During this time he is living in the midst of such a period of intellectual ferment as the world has seldom seen, and in a position which necessarily brings him into frequent contact with all the most cultivated men. Under such circumstances, there is nothing in the smallest degree strange or surprising in his acquiring the varied knowledge which his plays exhibit. The major premise of the Delia - Baconians has, therefore, nothing in it whatever. It is a mere bubble, an empty vagary, — only this, and nothing more.

Before leaving this part of the subject, however, there are still one or two points of interest to be mentioned. Shakespeare shows a fondness for the use of phrases and illustrations taken from the law ; and on such grounds our Delia-Baconians argue that the plays must have been written by an eminent lawyer, such as the Lord Chancellor Bacon undoubtedly was. They feel that this is a great point on their side. One instance, cited by Nathaniel Holmes and other Baconizers, is the celebrated case of Sir James Hales, who committed suicide by drowning, and was accordingly buried at the junction of crossroads, with a stake through his body, while all his property was forfeited to the Crown. Presently his widow brought suit for an estate by survivorship in joint-tenancy. Her case turned upon the question whether the forfeiture occurred during her late husband's lifetime : if it did, he left no estate which she could take; if it did not, she took the estate by survivorship. The lady's counsel argued that so long as Sir James was alive he had not been guilty of suicide, and the instant he died the estate vested in his widow as joint-tenant. But the opposing counsel argued that the instant Sir James voluntarily made the fatal plunge, and therefore before the breath had left his body, the guilt of suicide was incurred and the forfeiture took place. The court decided in favour of this view, and the widow got nothing.

There can be little doubt that this decision is travestied in the conversation of the two clowns in " Hamlet " with regard to Ophelia's right to Christian burial. The first clown makes precisely the point upon which the ingenious counsel for the defendant had rested his argument: " If I drown myself wittingly, it argues an act, and an act hath three branches ; it is to act, to do, and to perform." In making this distinction the counsel had maintained that the second branch, or the doing, was the only thing for the law to consider. The talk of the clowns brings out the humour of the case with Shakespeare's inimitable lightness of touch.

The report of the Hales case was published in the volume of "Plowden's Reports" which was issued in 1578; and Mr. Holmes informs us that "there is not the slightest ground for a belief, on the facts which we know, that Shakespeare ever looked into 'Plowden's Reports.'" This is one of the cases where your stern Baconizer will not hear of anything short of absolute demonstration. Mere considerations of human probability might disturb the cogency of a neat little pair of syllogisms : —

(1.) The author of "Hamlet" must have read Plowden. Shakespeare never read Plowden. Therefore Shakespeare was not the author of "Hamlet."

(2.) The author of "Hamlet" must have read Plowden. The lawyer, Bacon, must have read Plowden. Therefore Bacon wrote "Hamlet."

With regard to the major premise here, one may freely deny it. The author of "Hamlet" might easily have got all the knowledge involved from an evening chat with some legal friend at an alehouse. Then as to the minor premise, what earthly improbability is there in Shakespeare's having dipped into Plowden? Can nobody but lawyers or

law students enjoy reading **reports of law cases?**
I remember that, when I was about ten years old, a
favourite book with me was one entitled "Criminal
Trials of All Countries, by a Member of the Philadelphia Bar." I read it and read it, until forbid-
den to read such a gruesome book, and then I read
it all the more. One of the most elaborate reports
in it was that of the famous case of Captain Donellan, tried in 1780 on a charge of poisoning his
wife's brother, Sir Theodosius Boughton, a dis-
sipated and diseased young man, who died very
suddenly one day. A post-mortem inspection
showed spots in the intestine, which three ordinary
country doctors ascribed to poisoning by laurel
water, while Sir John Hunter, one of the greatest
authorities in Europe, testified that they might
equally well have ensued upon death from apo-
plexy. The judge, Sir Francis Buller, saw fit, in
his charge, to reckon this as the testimony of three
experts against one; and thus the jury were driven
to a verdict of murder, though it was not proved
that any murder had been committed. Captain
Donellan, who lived in his brother-in-law's house,
was a man of blameless life, an amateur chemist,
much given to fooling with odorous liquids and
hissing retorts. It was proved that he had been
distilling laurel water, and one or two other sus-

picious circumstances were alleged. The whole trial was begun and ended on the same day, the jury were about twenty minutes in finding the captain guilty, and three days afterward he was hung. It was a case where reason was submerged and drowned under a wave of angry prejudice shrieking for a victim.

Now, if I did not forthwith write a play, and take the occasion to ridicule the judge's charge to the jury, it was because I could not write a play, not because I did not fully appreciate the insult to law and common sense which that unfortunate case involved. In view of this and other experiences, when I now read a play or a novel that contains an intelligent allusion to some law case, I am far from feeling driven to the conclusion that it must have been written by a lord chancellor.

If Shakespeare's dramas are proved by such internal evidence to have been written by a lawyer, that lawyer, by parity of reasoning, could hardly have been Francis Bacon. For he was preëminently a chancery lawyer, and chancery phrases are in Shakespeare conspicuously absent. The word "injunctions" occurs five times in the plays, once perhaps with a reference to its legal use ("Merchant of Venice," II. ix.); but nowhere do we find any exhibition of a knowledge of chancery

law. His allusions to the common law are often
very amusing, as when, in " Love's Labour 's Lost,"
at the end of a brisk punning-match between Boyet
and Maria, he offers to kiss her, laughingly asking
for a grant of pasture on her lips, and she replies,
" Not so ; my lips are no common, though several
they be." Again, in " The Comedy of Errors,"
" Dromio asserts that there is no time for a bald
man to recover his hair. This having been written, the law phrase suggested itself, and he was
asked whether he might not do it by fine and recovery, and this suggested the efficiency of that
proceeding to bar heirs ; and this started the con-
ceit that thus the lost hair of another man would be
recovered." [1] In such quaint allusions to the common law and its proceedings Shakespeare abounds,
and we cannot help remembering that Nash, in his
prefatory epistle to Greene's " Menaphon," printed
about 1589, makes sneering mention of Shakespeare as a man who had left the " trade of Noverint," whereunto he was born, in order to try his
hand at tragedy. The " trade of Noverint " was
a slang expression for the business of attorney ;
and this passage has suggested that Shakespeare
may have spent some time in a law office, as student or as clerk, either before leaving Stratford, or

[1] Davis, *The Law in Shakespeare*, St. Paul, 1884.

perhaps soon after his arrival in London. This
seems to me not improbable. On the other hand,
"The Merchant of Venice" contains such crazy
law that it is hard to imagine it coming even from
a lawyer's clerk. At all events, we may safely say
that the legal knowledge exhibited in the plays is
no more than might readily have been acquired by
a man of assimilative genius associating with lawyers. It simply shows the range and accuracy of
Shakespeare's powers of observation.

Let us come now to the second part of the Delia
Bacon theory. Having satisfied herself that William Shakespeare could not have written the poems
and plays published under his name, she jumped
to the conclusion that Francis Bacon was the author. Surely, a singular choice! Of all men,
why Francis Bacon?[1] Why not, as I said before,
George Chapman or Ben Jonson, men who were
at once learned scholars and great poets? Chapman, like Marlowe, could write the "mighty line."
Jonson had rare lyric power; his verses sing, as
witness the wonderful "Do but look on her eyes,"
which Francis Bacon could no more have written
than he could have jumped over the moon. To

---

[1] There is reason for believing that this choice was an instance
of the megalomania developed by Miss Bacon's malady. She
imagined a remote kinship between herself and Lord Bacon.
Possibly there may have been such kinship.

pitch upon Bacon as the writer of " Twelfth Night "
or " Romeo and Juliet " is about as sensible as to
assert that " David Copperfield " must have been
written by Charles Darwin. After a familiar acquaintance of more than forty years with Shakespeare's works, of nearly forty years with Bacon's,
the two men impress me as simply antipodal one
to the other. A similar feeling was entertained
by the late Mr. Spedding, the biographer and editor of Bacon; and no one has more happily hit off
the vagaries of the Baconizers than the foremost
Bacon scholar now living, Dr. Kuno Fischer, in his
recent address before the Shakespeare Society at
Weimar.[1] I used to wonder whether the Bacon-
Shakespeare people really knew anything about
Bacon, and, now that chance has led me to read
their books, I am quite sure they do not. To their
minds, his works are simply a storehouse of texts
which serve them for controversial missiles, very
much as scattered texts from the Bible used to
serve our uncritical grandfathers.

Francis Bacon was one of the most interesting
persons of his time, and, as is often the case with
such many-sided characters, posterity has held various opinions about him. On the one hand, his
fame has grown brighter with the years; on the

[1] Fischer, *Shakespeare und die Bacon Mythen*, Heidelberg, 1895.

other hand, it has come to be more or less circumscribed and limited. Pope's famous verse, "The wisest, brightest, meanest of mankind," may be disputed in all its three specifications. Bacon's treatment of Essex, which formerly called forth such bitter condemnation, has been, I think, completely justified; and as for the taking of bribes, which led to his disgrace, there were circumstances which ought largely to mitigate the severity of our judgment. But if Bacon was far from being a mean example of human nature, it is surely an exaggeration to call him the wisest and brightest of mankind. He was a scholar and critic of vast accomplishments, a writer of noble English prose, and a philosopher who represented rather than inaugurated a most beneficial revolution in the aims and methods of scientific inquiry. He is one of the real glories of English literature, but he is also one of the most overrated men of modern times. When we find Macaulay saying that Bacon had "the most exquisitely constructed intellect that has ever been bestowed on any of the children of men," we need not be surprised to find that his elaborate essay on Bacon is as false in its fundamental conception as it is inaccurate in details. For a long time it was one of the accepted commonplaces that Bacon inaugurated the method by

which modern discoveries in physical science have
been made. Early in the present century, such
writers on the history of science as Whewell began
to show the incorrectness of this notion, and it was
completely exploded by Stanley Jevons in his
" Principles of Science," the most profound treatise
on method that has appeared in the last fifty years.
Jevons writes : " It is wholly a mistake to say that
modern science is the result of the Baconian philosophy ; it is the Newtonian philosophy and the
Newtonian method which have led to all the great
triumphs of physical science, and . . . the ' Prin-
cipia' forms the true Novum Organon." This
statement of Jevons is thoroughly sound. The
great Harvey, who knew how scientific discoveries
are made, said with gentle sarcasm that Bacon
" wrote philosophy like a lord chancellor ; " yet
Harvey would not have denied that the chancellor
was doing noble service as the eloquent expounder
of many sides of the scientific movement that was
then gathering strength. Bacon's mind was eminently sagacious and fertile in suggestions, but
the supreme creative faculty, the power to lead
men into new paths, was precisely the thing which
he did not possess. His place is a very high one
among intellects of the second order ; but to rank
him with such godlike spirits as Newton, Spinoza,

and Leibnitz simply shows that one has no real knowledge of the work which such men have done.

So much for Bacon himself. With regard to him as possible author of the Shakespeare poems and plays, it is difficult to imagine so learned a scholar making the kind of mistakes that abound in those writings. Bacon would hardly have introduced clocks into the Rome of Julius Cæsar; nor would he have made Hector quote Aristotle, nor Hamlet study at the University of Wittenberg, founded five hundred years after Hamlet's time; nor would he have put pistols into the age of Henry IV., nor cannon into the age of King John; and we may be pretty sure that he would not have made one of the characters in "King Lear" talk about Turks and Bedlam. In this severely realistic age of ours, writers are more on their guard against such anachronisms than they were in Shakespeare's time; in his works we cannot call them serious blemishes, for they do not affect the artistic character of the plays, but they are certainly such mistakes as a scholar like Bacon would not have committed.

Deeper down lies the contrast involved in the fact that Bacon was in a high degree a subjective writer, from whom you are perpetually getting revelations of his idiosyncrasies and moods, whereas

of all writers in the world Shakespeare is the most
completely objective, the most absorbed in the
work of creation. * In the one writer you are always reminded of the man Bacon; in the other
the personality is never thrust into sight. Bacon
is highly self-conscious; from Shakespeare self-
consciousness is absent.

The contrast is equally great in respect of
humour. I would not deny that Bacon relished
a joke, or could perpetrate a pun; but the bub-
bling, seething, frolicsome, irrepressible drollery of
Shakespeare is something quite foreign to him.
Read his essays, and you get charming English,
wide knowledge, deep thought, keen observation,
worldly wisdom, good humour, sweet serenity;
but exuberant fun is not there. In writing these
essays Bacon was following an example set by
Montaigne, but, as contrasted with the delicate effer-
vescent humour of the Frenchman, his style seems
sober and almost insipid. Only fancy such a man
trying to write " The Merry Wives of Windsor " !

Both Shakespeare and Bacon were sturdy and ra-
pacious purloiners. They seized upon other men's
bright thoughts and made them their own without
compunction and without acknowledgment; and
this may account for sundry similarities which may
be culled from the plays and from Bacon's works,

upon which Baconizing text-mongers are wont to lay great stress as proof of common authorship. Some such resemblances may be due to borrowing from common sources ; others are doubtless purely fanciful; others indicate either that Shakespeare cribbed from Bacon or *vice versa*. Here are a few miscellaneous instances : —

Where Bacon says, " Be so true to thyself as thou be not false to others "(" Essay of Wisdom "), Shakespeare says : —

> " To thine own self be true,
> And it must follow, as the night the day,
> Thou canst not then be false to any man."
>
> (Hamlet, I. iii.)

This looks as if one writer might have copied from the other. If so, it is Bacon who is the thief, for the lines occur in the quarto " Hamlet " published in 1603, whereas the " Essay of Wisdom " was first published in 1612.

Again, where Bacon, in the " Essay of Gardens," says, " The breath of flowers comes and goes like the warbling of music," it reminds one strongly of the exquisite passage in " Twelfth Night " where the Duke exclaims : —

> " That strain again ! it had a dying fall:
> O, it came o'er my ear like the sweet south,
> That breathes upon a bank of violets,
> Stealing and giving odour."

I have little doubt that Bacon had this passage in mind when he wrote the " Essay of Gardens," which was first published in 1625, two years later than the complete folio of Shakespeare. This effectually disposes of the attempt to cite these correspondences in evidence that Bacon wrote the plays.

Another instance is from " Richard III. : " —

> " By a divine instinct men's minds mistrust
> Ensuing danger ; as, by proof, we see
> The waters swell before a boisterous storm."

Bacon, in the " Essay of Sedition," writes, " As there are . . . secret swellings of seas before a tempest, so there are in states." But this essay was not published till 1625, so again we find him copying Shakespeare. Many such " parallelisms," cited to prove that Bacon wrote Shakespeare's works, do really prove that he read them with great care and remembered them well, or else took notes from them.

An interesting illustration of the helpless ignorance shown by Baconizers is furnished by a remark of Sir Toby Belch in " Twelfth Night." In his instructions to that dear old simpleton, Sir Andrew Aguecheek, about the challenge, Sir Toby observes, " If thou thou'st him some thrice, it shall not be amiss." In Elizabethan English, to address a man as "thou" was to treat him as socially inferior; such familiarity was allowable only between members of the same family or in speaking to servants, just as you address your wife, and likewise the cook and housemaid, by their Christian names, while with the ladies of your acquaintance such familiarity would be rudeness. The same rule for the pronoun survives to-day in French and German, but has been forgotten in English. In the trial of Sir Walter Raleigh in 1604, Justice Coke insulted the prisoner by calling out, "Thou viper! for I *thou* thee, thou traitor!" Now, one of our Baconizers thinks that his idol, in writing "Twelfth Night," introduced Sir Toby's suggestion in order to recall to the audience Coke's abusive remark. Once more, a little attention to dates would have prevented the making of a bad blunder. We know from Manningham's Diary that "Twelfth Night" had been on the stage nearly two years before Raleigh's trial. On the other hand, to say that the play might have suggested to Coke his coarse speech would be admissible, but idle, inasmuch as the expression "to *thou* a man" was an every-day phrase in that age.

Here it naturally occurs to me to mention the "Promus," about which as much fuss has been made as if it really furnished evidence in support of the Baconian folly. There is in the British Museum a manuscript, in Bacon's handwriting, entitled " Promus of Formularies and Elegancies." " Promus " means " storehouse " or " treasury." A date at the top of the first page shows that it was begun in December, 1594 ; there is nothing, I believe, to show over how many years it extended. It is a scrap-book in which Bacon jotted down such sentences, words, and phrases as struck his fancy, such as might be utilized in his writings. These neatly turned phrases, these " formularies and elegancies," are gathered from all quarters, — from the Bible, from Virgil and Horace, from Ovid and Seneca, from Erasmus, from collections of proverbs in various languages, etc. As there is apparently nothing original in this scrap-bag, Mr. Spedding did not think it worth while to include it in his edition of Bacon's works, but in the fourteenth volume he gives a sufficient description of it, with illustrative extracts. In 1883 Mrs. Henry Pott published the whole of this " Promus " manuscript, and swelled it by comments and dissertations into a volume of 600 octavo pages. She had found in it several hundred expressions which reminded her of passages in Shakespeare, and so it confirmed her in the opinion which she already entertained that Bacon was the author of Shake-

speare's works. Thus, when the "Promus" has a
verse from Ovid, which means, "And the forced
tongue begins to lisp the sound commanded," it
reminds Mrs. Pott of divers lines in which Shakespeare uses the word "lisp," as for example, in
"As You Like It," "you lisp and wear strange
suits;" and she jumps to the conclusion that when
Bacon jotted down the verse from Ovid, it was as
a preparatory study toward "As You Like It,"
and any other play that contains the word "lisp:"
therefore Bacon wrote all those plays, *Q. E. D.!*
On the next page we find Virgil's remark, "Thus
was I wont to compare great things with small,"
made the father of Falstaff's "base comparisons,"
and Fluellen's "Macedon and Monmouth," as well
as honest Dogberry's "comparisons are odorous."
When one reads such things, evidently printed in
all seriousness, one feels like asking Mrs. Pott,
in the apt words of Shakespeare's friend Fletcher,
"What mare's nest hast thou found?" ("Bon-
duca," V. ii.)

There are many phrases, however, in the "Promus" which undoubtedly agree with phrases in the
plays. They show that Bacon heard or read the
plays with great interest, and culled from them his
"elegancies" with no stinted hand. As for Mrs.
Pott's bulky volume, it brings us so near to the final

reductio ad absurdum of the Bacon theory that we hardly need spend many words upon the gross improbabilities which that theory involves. The plays of Shakespeare were universally ascribed to him by his contemporaries; many of them were published during his lifetime with his name upon the title-page as the author; all were collected and published together by Hemminge and Condell, two of his fellow actors, seven years after his death; and for more than two centuries nobody ever dreamed of looking for a different authorship, or of associating the plays with Bacon. But this Chimborazo of *prima facie* evidence becomes a mere mole-hill in the hands of your valiant Baconizer. It is all clear to him. Bacon did not acknowledge the authorship of these works because such literature was deemed frivolous, and current prejudices against theatres and playwrights might injure his hopes of advancement at the bar and in political life. Therefore, by some sort of private understanding with the ignorant and sordid wretch Shakespeare,[1] at whose theatre they were brought out, their authorship was ascribed to him, the real author died without revealing the secret, and the whole world was deceived until the days of Delia Bacon.

[1] The Baconizers usually delight in berating poor Shakespeare, making much of the deer-stealing business, the circumstances of his marriage, etc.

But there are questions which even this ingenious hypothesis fails to answer. Why should Bacon have taken the time to write those thirty-seven plays, two poems, and one hundred and fifty-four sonnets, if they were never to be known as his works? Not for money, surely, for that grasping Shakespeare seems to have got the money as well as the fame; Bacon died a poor man. His principal aim in life was to construct a new system of philosophy; on this noble undertaking he spent such time as he could save from the exactions of his public career as member of Parliament, chancery lawyer, solicitor-general, attorney-general, lord chancellor; and he died with this work far from finished. The volumes which he left behind him were only fragments of the mighty structure which he had planned. We may well ask, Where did this overburdened writer find the time for doing work of another kind voluminous enough to fill a lifetime, and what motive had he for doing it without recompense in either fame or money? Baconizers find it strange that Shakespeare's will contains no reference to his plays as literary property. The omission is certainly interesting, since it seems to indicate that he had parted with his pecuniary interest in them, — had perhaps sold it out to the Globe Theatre. If this omission can be held to show that Shakespeare was lacking in fondness for the productions of his own genius, what shall be said of the notion that Bacon spent half his life in writing works the paternity of which he must forever disown?

This question is answered by Mr. Ignatius Donnelly, a writer who speculates with equal infelicity on all subjects, but never suffers for lack of boldness. He published in 1887 a book even bigger than that of Mrs. Pott, for it has nearly 1000 pages. Its title is, "The Great Cryptogram," and its thesis is, that Bacon really did claim the authorship of the Shakespeare plays. Only the claim was made in a cipher, and if you simply make some numbers mean some words, and other words mean other numbers, and perform a good many sums in what the Mock Turtle called "ambition, distraction, uglification, and derision," you will be able to read this claim between the lines, along with much other wonderful information. Thus does the arithmetical Donnelly carry us quite a long stride nearer to the *reductio ad absurdum*, or suicide point, than we were left by Mrs. Pott, with her lisping and limping comparisons.

But before we come to the jumping-off place, let us pause for a moment and take a retrospective glance at the natural history of the Bacon-Shake-

speare craze.   What was it that first unlocked
the sluice-gates, and poured forth such a deluge
of foolishness upon a sorely suffering world?   It
will hardly do to lay the blame upon poor Delia
Bacon.   Her suggestions would have borne no fruit
had they not found a public, albeit a narrow one,
in some degree prepared for them.   Who, then,
prepared the soil for the seeds of this idiocy to
take root?   Who but the race of fond and fool-
ish Shakespeare commentators, with their absurd
claims for their idol ?   During the eighteenth century Shakespeare was generally underrated.   Voltaire wondered how a nation that possessed such a
noble tragedy as Addison's " Cato " could endure
such plays as " Hamlet " and " Othello."   In the
days of Scott and Burns a reaction set in ; and
Shakespeare worship reached its height when the
Germans took it up, and, not satisfied with calling
him the prince of poets and peerless master of
dramatic art, began to discover in his works all
sorts of hidden philosophy and impossible knowledge.   Of the average German mind Lowell good-
naturedly says that " it finds its keenest pleasure
in divining a profound significance in the most
trifling things, and the number of mare's nests
that have been stared into by the German *Gelehr-
ter* through his spectacles passes calculation." [1]

[1] *Literary Essays*, ii. 163.

But the Germans are not the only sinners ; let me cite an instance from near home. In the quarto " Hamlet " of 1603 we read : —

> " Full forty years are past, their date is gone,
> Since happy time joined both our hearts as one :
> And now the blood that filled my youthful veins
> Runs weakly in their pipes," etc.

Whereupon Mr. Edward Vining calls upon us to observe how Shakespeare, " to whom all human knowledge seems to be but a matter of instinct, in [these lines] asserts the circulation of the blood in the veins and ' pipes,' a truth which Harvey probably did not even suspect until at least thirteen years later," etc.[1] Does Mr. Vining really suppose that what Harvey did was to discover that blood runs in our veins? A little further study of history would have taught him that even the ancients knew that blood runs in the veins.[2] About fourteen hundred years before " Hamlet " was written, Galen proved that it also runs in the

---

[1] The Bankside *Shakespeare*, vol. xi. p. xi.

[2] The writings of Hippocrates abound in examples, as in his interesting explanation of congestion, extravasation, etc. (*De Ventis*, x.-xiv., *Opera*, ed. Littré, tom. vi. pp. 104-114), to cite one instance out of a thousand : Ἐπειδὰν οὖν ἐς τὰς παχείας καὶ πολυαίμους τῶν φλεβῶν πολὺς ἀὴρ βρίσῃ, βρίσας δὲ μένῃ, κωλύεται τὸ αἷμα διεξιέναι· τῇ μὲν οὖν ἐνέστηκε, τῇ δὲ νωθρῶς διεξέρχεται, τῇ δὲ θᾶσσον· etc.

arteries. After Galen's time, it was believed that
the dark blood nourishes such plebeian organs as
the liver, while the bright blood nourishes such
lordly organs as the brain, and that the interchange takes place in the heart; until the sixteenth century, when Vesalius proved that the
interchange does not take place in the heart, and
the martyr Servetus proved that it does take place
in the lungs; and so on till 1619, when Harvey
discovered that dark blood is brought by the veins
to the right side of the heart, and thence driven
into the lungs, where it becomes bright and flows
into the left side of the heart, thence to be propelled throughout the body in the arteries. That
it then grows dark and returns through the veins
Harvey believed, but no one could tell how, until,
forty years later, Malpighi with his microscope detected the capillaries. Now to talk about Shakespeare discerning as if by instinct a truth which
Harvey afterward discovered is simply silly. Instead of showing rare scientific knowledge, his
remark about blood running in the veins is one
that anybody might have made.

This is a fair specimen of the ignorant way
in which doting commentators have built up an
impossible Shakespeare, until at last they have
provoked a reaction. Sooner or later the question

was sure to arise, Where did your Stratford boy get all this abstruse scientific knowledge? The keynote was perhaps first sounded by August von Schlegel, who persuaded himself that Shakespeare had mastered " all the things and relations of this world," and then went on to declare that the accepted account of his life must be a mere fable. Thus we reach the point from which Delia Bacon started.

It may safely be said that all theories of Shakespeare's plays which suppose them to be attempts at teaching occult philosophical doctrines, or which endow them with any other meanings than those which their words directly and plainly convey, are a delusion and a snare. Those plays were written, not to teach philosophy, but to fill the theatre and make money. They were written by a practised actor and manager, the most consummate master of dramatic effects that ever lived; a poet unsurpassed for fertility of invention, unequalled for melody of language, unapproached for delicacy of fancy, inexhaustible in humour, profoundest of moralists; a man who knew human nature by intuition, as Mozart knew counterpoint or as Chopin knew harmony. The name of that writer was none other than William Shakespeare of Stratford-on-Avon.

It was inevitable that the Bacon folly, after once adopting such methods as those of Mrs. Pott and Mr. Donnelly, should proceed to commit suicide by piling up extravagances. By such methods one can prove anything, and accordingly we find these writers busy in tracing Bacon's hand in the writings of Greene, Marlowe, Shirley, Marston, Massinger, Middleton, and Webster. They are sure that he was the author of Montaigne's Essays, which were afterward translated into what we have always supposed to be the French original. Mr. Donnelly believes that Bacon also wrote Burton's "Anatomy of Melancholy." Next comes Dr. Orville Owen with a new cipher, which proves that Bacon was the son of Queen Elizabeth by Robert Dudley, and that he was the author of the "Faerie Queene" and other poems attributed to Edmund Spenser. Finally we have Mr. J. E. Roe, who does not mean to be outdone. He asks us what we are to think of the notion that an ignorant tinker, like John Bunyan, could have written the most perfect allegory in any language. Perish the thought! Nobody but Bacon could have done it. Of course Bacon had been more than fifty years in his grave when "Pilgrim's Progress" was published as Bunyan's. But your true Baconizer is never stopped by trifles. Mr. Roe assures us that

Bacon wrote that heavenly book, as well as " Robinson Crusoe " and the " Tale of a Tub ; " which surely begins to make him seem ubiquitous and everlasting. If things go on at this rate, we shall presently have a religious sect holding as its first article of faith that Francis Bacon created the heavens and the earth in six days, and rested on the seventh day.

November, 1896.

XIV

SOME CRANKS AND THEIR CROTCHETS

" Now, by two-headed Janus,
Nature hath framed strange fellows in her time ! "
Merchant of Venice, I. i.

ABOUT five-and-twenty years ago, when I was assistant librarian at Harvard University, much of my time was occupied in revising and bringing toward completion the gigantic pair of twin catalogues — of authors and subjects — which my predecessor, Dr. Ezra Abbot, had started in 1861. Twins they were in simultaneity of birth, but not in likeness of growth. Naturally, the classified catalogue was much bigger than its brother, filled more drawers, cost more money, and made a vast deal more trouble. For while some books were easy enough to classify, others were not at all easy, and sometimes curious questions would arise.

One day, for example, I happened to be looking at a pamphlet on the value of Pi; and, should any of my readers ask what that might mean, I should answer that Pi (π) is the Greek letter which geometers use to denote the ratio of the circumference of a circle to its diameter. The arithmetical value of this symbol is 3.1415926536, and so on in an endless fraction. Is it not hard to see what there can be in such an innocent decimal to irritate human beings and destroy their peace of mind? Yet so it is. Many a human life has been wrecked upon Pi. To a certain class of our fellow-creatures its existence is maddening. It interferes with the success of a little scheme on which they have set their hearts, — nothing less than to construct a square which shall be exactly equivalent in dimensions to a given circle. Nobody has ever done such a thing, for it cannot be done. But when mathematicians tell these poor people that such is the case, they howl with rage, and, dipping their pens in gall, write book after book bristling with figures to prove that they have " squared the circle." The Harvard library does not buy such books, but it accepts all manner of gifts, and has thus come to contain some queer things.

When I consulted the subject catalogue, to see under what head it had been customary to classify these lucubrations on Pi, I found, sure enough, that it was Mathematics § Circle-Squaring. Following this cue, I explored the drawers in other directions, and found that books on " perpetual

motion" formed a section under Physics, while
crazy interpretations of the book of Daniel were
grouped along with works of solid Biblical scholarship by such eminent writers as Reuss and Kuenen
and Cheyne. Clearly, here was a case for reform.
The principle of classification was faulty. In one
sense, the treatment of the quadrature of the circle
may be regarded as a section under the general
head of mathematics; as, for example, when Lin-
demann, in 1882, showed that Pi cannot be represented as the root of any algebraic equation with
rational coefficients. But our circle-squaring literature is very different. It is usually written by
persons whose mathematical horizon scarcely extends beyond long division: just as the writers on
perpetual motion know nothing of physics; just as
so many expositors have dealt with the ten-horned
beast in blissful ignorance alike of ancient history
and of the principles of literary criticism. What
all such books illustrate, however various may be
their ostensible themes, is the pathology of the
human mind. They are specimens of Insane Literature. As such they have a certain sort of interest; and to any rational being it is the only sort
they can have.

So I culled from many a little drawer the cards
appertaining to divers printed products of morbid

cerebration, and gathered them into a class of Insane Literature ; and under this rubric such sections as Circle-Squaring, Perpetual Motion, Great Pyramid, Earth not a Globe, etc., were evidently in their proper place. The name of the class was duly inscribed on the outside of its drawer, and the matter seemed happily disposed of.

The way of the reformer, however, is beset with difficulties, and it is seldom that his first efforts are crowned with entire success. Not many days had elapsed since this emendation of the catalogue, when one of my assistants brought me the card of a book on the Apocalypse, by a certain Mr. Smallwit, and called my attention to the fact that it was classified as Insane Literature.

" Very well," I said, " so it is." ·

"I don't doubt it, sir," said she; " but the author lives over in Chelsea, and I saw him this morning in one of the alcoves. Perhaps, if he were to look in the catalogue and see how his book is classified, he might n't altogether like it. Then, as I looked a little further along the cards, I came upon this pamphlet by Herr Dummkopf, of Breslau, upsetting the law of gravitation ; and — do you know? — Herr Dummkopf is spending the winter here in Cambridge ! "

" To be sure," said I, " it was very stupid of me

not to foresee such cases. Of course we can't call a man a fool to his face. In a catalogue which marshals the quick along with the dead some heed must be paid to the amenities of life. Pray get and bring me all those cards."

By the time they arrived a satisfactory solution of the difficulty had suggested itself. I told the assistant simply to scratch out "Insane," and put "Eccentric" instead. For while the harsh Latin epithet would of course infuriate Messrs. Dummi-kopf, Smallwit & Co., it might be doubted if their feelings would be hurt by the milder Greek word. Some people of their stripe, to whom notoriety is the very breath of their nostrils, would consider it a mark of distinction to be called eccentric. At all events, the harshness would be delicately veiled under a penumbra of ambiguity.

Thus the class Eccentric Literature was established in our catalogue, and there it has remained, while the books in the library have increased from a hundred thousand to half a million. Once or twice, I am told, has some disgusted author uttered a protest, but the quiet of Gore Hall has not been disturbed thereby. Care is needed in treating such a subject, and my rule was that no amount of mere absurdity, no extremity of dissent from generally received opinions, should consign a book to the class of Eccentric Literature, unless it showed unmistakable symptoms of crankery, or the buzzing of a bee in the author's bonnet. This rule has been strictly followed. One lot of books — the Bacon-Shakespeare stuff — which I intended to put in this class, but forgot to do so because of sore stress of work, still remain absurdly grouped along with the books on Shakespeare written by men in their senses. With this exception, the class offers us a fairly comprehensive view of the literature of cranks.

Just where the line should be drawn between sanity and crankery is not always easy to determine, and must usually be left to soundness of judgment in each particular case, as with so many other questions of all grades, from the supreme court down to the kitchen. One of the most frequent traits of your crank is his megalomania, or self-magnification. His intellectual equipment is so slender that he cannot see wherein he is inferior to Descartes or Newton. Without enough knowledge to place him in the sixth form of a grammar school, he will assail the conclusions of the greatest minds the world has seen. His mood is belligerent; since people will not take him at his own valuation, he is apt to regard society as engaged in a conspiracy to ignore and belittle him.

Of humour he is pretty sure to be destitute; an abounding sense of the ludicrous is one of the best safeguards of mental health, and even a slight endowment will usually nip and stunt the fungus growth of crankery.

The slightest glimmering sense of humour would have restrained that inveterate circle-squarer, James Smith, from publishing (in 1865) his pamphlet entitled "The British Association in Jeopardy, and Dr. Whewell, the Master of Trinity, in the Stocks without Hope of Escape." His case, with those of many other ingenious lunatics, was racily set forth by the late Professor De Morgan in his "Budget of Paradoxes" (London, 1872), a bulky book dealing with the author's personal experiences with cranks and their crotchets. It was De Morgan's lot as an eminent mathematician to be outrageously bored by circle-squarers and their kin, and it was a happy thought to put on record the queer things that happened. His friends asked him again and again why he took the trouble to mention and expose such absurdities. He replied that, when your crank publishes a book "full of figures which few readers can criticise, a great many people are staggered to this extent, that they imagine there must be the indefinite *something* in the mysterious *all this*. They are brought to the point of suspicion that the mathematicians ought not to treat *all this* with such undisguised contempt, at least. Now I have no fear for π; but I do think it possible that general opinion might in time demand that the crowd of circle-squarers, etc., should be admitted to the honours of opposition; and this would be a time-tax of five per cent, one man with another, upon those who are better employed." At any rate, continues De Morgan, with a twinkle in the corner of his eye, whether in chastising cranks he has any motive but public good "must be referred to those who can decide whether a missionary chooses his pursuit solely to convert the heathen." He confesses that perhaps he may have a little of the spirit of Colonel Quagg, whose principle of action was thus succinctly expressed: "I licks ye because I kin, and because I like, and because ye's critters that licks is good for!"

Among the creatures whose malady seemed to call for such drastic treatment was Captain Forman, R. N., who in 1833 wrote against the law of gravitation, and got not a word of notice. Then he wrote to Sir John Herschel and Lord Brougham, asking them to get his book reviewed in some of the quarterlies. Receiving no answer from these gentlemen, he addressed in one of the newspapers a card to Lord John Russell, inveighing against their "dishonest" behaviour. Still getting no satisfaction, the valorous captain wrote to the Royal Astronomical Society with a challenge to controversy. To this letter came a polite but brief answer, advising him to study the rudiments of mechanics. It was not in the paradoxer's nature to submit tamely to such treatment; and he replied in a printed pamphlet, wherein he called that learned society "craven dunghill cocks," and bestrewed them with other choice flowers of rhetoric, much to the relief of his feelings.

One of this naval officer's fellow sufferers was a farm labourer, who took it into his head that the Lord Chancellor had offered £100,000 reward to any one who should square the circle. So Hodge went to work and squared it, and then hied him to London, blissfully dreaming of sudden wealth. Hearing that De Morgan was a great mathematician, he left his papers with him, including a letter to the Lord Chancellor, claiming the £100,000. De Morgan returned the papers with a note, saying that no such prize had ever been offered, and gently hinting that the worthy Hodge had not sufficient knowledge to see in what the problem consisted. This elicited from the rustic philosopher a long letter, from which I must quote a few sentences, so characteristic of the circle-squaring talent and temper : —

Doctor Morgan, Sir. Permit me to address you

Brute Creation may perhaps enjoy the faculty of beholding visible things with a more penitrating eye than ourselves. But Spiritual objects are as far out of their reach as though they had no being Nearest therefore to the brute Creation are those men who Suppose themselves to be so far governed by external objects as to believe nothing but what they See and feel And Can accomedate to their Shallow understanding and Imaginations

. . . When a Gentleman of your Standing in Society . . . Can not understand or Solve a problem That is explicitly explained by words and Letters and mathematacally operated by figuers He had best consult the wise proverd

Do that which thou Canst understand and Comprehend for thy good.

I would recommend that Such Gentleman Change his business

And appropriate his time and attention to a Sunday School to Learn what he Could and keep the Litle Children form durting their Close

With Sincere feelings of Gratitude for your weakness and Inability I am

Sir your Superior in Mathematics.

X. Y.

A few days after this elegant epistle there came
to De Morgan another from the same hand.
Hodge had sent his papers to some easy-going
American professor, whose reply must clearly have
been too polite. It is never safe to give your
crank an inch of comfort; it will straightway become an ell of assurance. This American savant,
crows Rusticus, " highly approves of my work.
And Says he will Insure me Reward in the States
I write this that you may understand that I have
knowledge of the unfair way that I am treated in
my own nati County I am told and have reasons
to believe that it is the Clergy that treat me so
unjust. I am not Desirious of heaping Disonors
upon my own nation. But if I have to Leave this
kingdom without my Just dues. The world Shall
know how I am and have been treated

" I am Sir Desirous of my Just dues

" X. Y."

A cynical philosopher once said that you cannot find so big a fool but there will be some bigger
fool to swear by him; and so our agricultural
friend had his admiring disciple who felt bound to
break a lance for him with the unappreciative De
Morgan : —

" He has done what you nor any other mathematician as those who call themselves such have done.

And what is the reason that you will not candidly acknowledge to him . . . that he has squared the circle shall I tell you? it is because he has performed the feat to obtain the glory of which mathematicians have battled from time immemorial that they might encircle their brows with a wreath of laurels far more glorious than ever conqueror won it is simply this that it is a poor man a humble artisan who has gained that victory that you don't like to acknowledge it you don't like to be beaten and worse to acknowledge that you have miscalculated, you have in short too small a soul to acknowledge that he is right. . . . I am backed in my opinion not only by Mr. Q. a mathematician and watchmaker residing in the boro of Southwark but by no less an authority than the Professor of mathematics of . . . College United States Mr. Q and I presume that he at least is your equal as an authority and Mr. Q says that the government of the U. S. will recompense X. Y. for the discovery he has made if so what a reflection upon Old england the boasted land of freedom the nursery of the arts and sciences that her sons are obliged to go to a foreign country to obtain that recompense to which they are justly entitled." [1]

Ordinarily, the aim of the paradoxers is to achieve

<hr>

[1] *Budget of Paradoxes,* pp. 9, 178, 259, 260, 336.

renown by doing what nobody ever did. Hence
the fascination exercised upon them by those apparently simple problems which already in ancient
times were recognized as "old stickers," the quadrature of the circle, the trisection of angles, and
the duplicature of the cube. The ancients found
these geometric problems insolvable, though it was
left for modern algebra to point out the reason,
namely, that no quantities can be geometrically
constructed from given quantities, except such as
can be formed from them algebraically by the solution of quadratic equations; if the algebraic solution
comes as the root of a cubic or biquadratic equation,
it cannot be constructed by geometry. Against
this hopeless wall the crowd of paradoxers will
doubtless continue to break their heads until the
millennium dawns.

Sometimes, however, our crank has a practical
end in view, as in the numerous attempts to discover "perpetual motion," or, in other words, to
invent a machine out of which you can get indefi-
nitely more energy than you put in. It is not
strange that many thousands of dollars have been
wasted in this effort to recover Aladdin's lost lamp.
The notorious Keely motor is but one of a host of
contrivances born and bred of crass ignorance of
the alphabet of dynamics. But perpetual motion

is not the only form assumed by wealth-seeking
crankery. In 1861 a Captain Roblin, of Normandy, having ascertained to his own satisfaction,
from the prolonged study of the zodiac of Dende-
rah, the sites of sundry gold-mines, came forward
with proposals for a joint stock company to dig
and be rich. The labours of Herr Johannes von
Gumpach were of a more philanthropic turn. He
published in 1861 a pamphlet entitled " A Million's Worth of Property and Five Hundred Lives
annually lost at Sea by the Theory of Gravitation.
A Letter on the True Figure of the Earth, addressed to the Astronomer Royal." Next year
this pamphlet grew into a stout volume. It maintained that a great many shipwrecks were occa-
sioned by errors of navigation due to an erroneous
conception of the shape of the earth. Since Newton's time, it has been supposed to be flattened at
the poles, whereas the amiable Gumpach calls upon
his fellow-creatures to take notice that it is elon-
gated, and to mend their ways accordingly.

The desire to prove great men wrong is one of
the crank's most frequent and powerful incentives.
The name of Newton is the greatest in the history
of science : how flattering to one's self it must be,
then, to prove him a fool ! In eccentric literature
the books against Newton are legion. Here is a

title: " David and Goliath, or an Attempt to prove
that the Newtonian System of Astronomy is
directly opposed to the Scriptures. By William
Lander, Mere, Wilts, 1833." And here is De
Morgan's terse summary of the book: " Newton
is Goliath; Mr. Lander is David. David took
five pebbles; Mr. Lander takes five arguments.
He expects opposition; for Paul and Jesus both
met with it."

There are few subjects over which cranks are
more painfully exercised than the figure of the
earth, and its relations to heavenly bodies. Aristotle proved that the earth is a globe; Copernicus
showed that it is one of a system of planets revolv-
ing about the sun; Newton explained the dynamics
of this system. But at length came a certain
John Hampden, who with dauntless breast maintained that all this is wrong! His pamphlet was
prudently dedicated "to the unprofessional public and the common sense men of Europe and
America; " he knew that it could find no favor
with bigoted men of science. This Hampden, like
his great namesake, is nothing if not bold. "The
Newtonian or Copernican theory," he · tells us,
"from the first hour of its invention, has never
dared to submit to an appeal to facts!" Again,
"Defenders it never had; and no threats, no

taunts or exposure, will ever rouse the energies of
a single champion." In other words, astronomers
do not waste their time in noticing Mr. Hampden's
taunts and threats. Why is this so? His next
sentence reminds us that " cowardice always accompanies conscious guilt." He goes on to tell
us the true state of the case: "The earth, as it
came from the hands of its Almighty Creator, is a
motionless Plane, based and built upon foundations
which the Word of God expressly declares cannot
be searched out or discovered. . . . The stars are
hardly bigger than the gas jets which light our
streets, and, if they could be made to change places
with them, no astronomer could detect the difference." The North Pole is the centre of the flat
earth, and its extreme southern limit is not a
South Pole, but a circle 30,000 miles in circumference. Night is caused by the sun passing behind a layer of clouds 7000 miles thick. It is not
gravitation which makes a river run down hill,
but the impetus of the water behind pressing on
the water before. Is not this delicious? As for
Newton, poor fellow, he " lived in a superstitious
age and district; he was educated among an illiterate peasantry." This is like the way in which
the Baconizing cranks dispose of Shakespeare. So
zealous was Mr. Hampden that in 1876 he began

publishing a periodical called " The Truth Seeker's Oracle." Similar views were set forth by one Samuel Rowbotham, who wrote under the name of " Parallax," and by a William Carpenter, whose pamphlet, "One Hundred Proofs that the Earth is not a Globe " (Baltimore, 1885), is quite a curiosity ; for example, Proof 33 : " If the earth were a globe, people — except those on top — would certainly have to be fastened to its surface by some means or other ; . . . but as we know that we simply walk on its surface, without any other aid than that which is necessary for locomotion on a plane, it follows that we have herein a conclusive proof that Earth is not a globe." Since Mr. Carpenter understands the matter so thoroughly, can we wonder at the earnestness with which he rebukes the late Richard Proctor ? " Mr. Proctor, we charge you that, whilst you teach the theory of the earth's rotundity, you KNOW that it is a plane ! "

More original than Messrs. Hampden and Carpenter are the writers who maintain that the earth is hollow, and supports a teeming population in its interior. Early in the present century this idea came with the force of a revelation to the mind of Captain John Cleves Symmes, a retired army officer engaged in trade at St. Louis. In 1818 he issued a circular, of which the following is an abridgment : " To ALL THE WORLD I declare the earth is hollow and habitable within ; containing a number of solid concentric spheres, one within the other, and that it is open at the poles twelve or sixteen degrees. I pledge my life in support of this truth, and am ready to explore the hollow, if the world will support and aid me in the undertaking. . . . My terms are [Hear, Messrs. Quay and Platt ! and give ear, O Tammany !] *the* PATRONAGE *of* THIS *and the* NEW WORLDS. . . . I select Dr. S. L. Mitchell, Sir H. Davy, and Baron Alexander von Humboldt as my protectors. I ask one hundred brave companions, well equipped, to start from Siberia, in the fall season, with reindeer and sleighs, on the ice of the frozen sea. I engage we find a warm and rich land, stocked with thrifty vegetables and animals, if not men, on reaching one degree northward of latitude 82°. We will return in the succeeding spring."

This circular was sent by mail to men of science, colleges, learned societies, legislatures, and municipal bodies, all over the United States and Europe ; for when it comes to postage, your crank seems always to have unlimited funds at his disposal. At Paris, the distinguished traveller, Count Volney, doubtless with a significant shrug, pre-

sented the precious document to the Academy of
Sciences, by which it was mirthfully laid upon the
table. Nowhere did learned men take it seriously;
it was generally set down as a rather stupid hoax.
But, nothing daunted by such treatment, the
worthy Symmes began giving lectures on the subject, and succeeded in making some impression
upon an uninstructed public. In 1824 his audience at Hamilton, Ohio, at the close of a lecture,
" *resolved*, that we esteem Symmes' Theory of the
Earth deserving of serious examination and worthy
of the attention of the American people." At a
theatre in Cincinnati, a benefit was given for the
proposed polar expedition, and verses were recited
suitable to the occasion : —

> " Has not Columbia one aspiring son
> By whom the unfading laurel may be won?
> Yes! history's pen may yet inscribe the name
> Of Symmes to grace her future scroll of fame."

The captain's petitions to Congress, however, pray-
ing for ships and men, were heartlessly laid on the
table, and nothing was left him but to keep on cry-
ing in the wilderness, which he did until his death
in 1829. In the cemetery at Hamilton, the freestone monument over his grave, placed there by
his son, Americus Symmes, is surmounted with a
hollow globe, open at the poles.

Half a century later the son published a pamphlet,[1] in which he gave a somewhat detailed exposition of his father's notions. From this we learn that the interior world is well lighted; for the sun's rays, passing through " the dense cold air of the verges " (that is, the circular edge of the big polar hole), are powerfully refracted, and after getting inside they are forthwith reflected from one concave surface to another, with the result that the whole interior is illuminated with a light equal to 3600 times that of the full moon. We learn, too, that the famous Swedish geographer, Norpensjould (*semper sic !*), after passing the magnetic pole, found a timbered country with large rivers and abundant animal life. Afterward one Captain Wiggins visited this country, where he found flax and wheat, highly magnetic iron ore, and rich mines of copper and gold. The trees are as big as any in California; hides, wool, tallow, ivory, and furs abound. The inhabitants are very tall, with Roman noses, and speak Hebrew. Yes, echoes Captain Tuttle, an old whaler, who also has visited this new country, they speak Hebrew, and are a smart people. " Would it not be logical," writes Americus, " to think that this was one of

[1] *The Theory of Concentric Spheres*, Louisville, 1878; second edition, 1885.

the lost tribes of Israel? for we read in the Bible that they went up the Euphrates to the north and dwelt in a land where man never dwelt before." Just so; evidently, Messrs. " Norpensjould," Wiggins, and Tuttle sailed "across the verge" and into the interior country, the concave world, which shall henceforth be known as Symmzonia! The book ends with the triumphant query, " Where were those explorers if not in the Hollow of the Earth, and would they not have come out at the South Pole if they had continued on their course?"

It is sad to have such positive conclusions disputed, but even in eccentric lore the doctors are found to disagree. Scarcely had Americus put forth his revised edition, when a pamphlet entitled "The Inner World," by Frederick Culmer, was published at Salt Lake City (1886). Its chapters have resounding titles: " I. The Universal Vacuity of Centres; II. The Polar Orifices of the Earth; III. The Alleged Northwest Passage and Symmes' Hole." We are told that although the polar orifices have diameters of about a thousand miles each, nevertheless, in spite of Wiggins and Tuttle, "there is no passage to the inner world on the north of America;" on the contrary, it must be sought within the antarctic circle. But Mr. Culmer would discourage rash attempts at exploration, and believes that " no man will be able to plant the standard of his country on any land in that region worth one dime to himself or any one else at present." For this gloomy outlook we must try to console ourselves with the knowledge that Mr. Culmer has detected the true explanation of the Aurora Borealis : " It is the sun's rays shining on a placid interior ocean and reflecting upon the outer atmosphere."

A favourite occupation of cranks is the discovery of hidden meanings in things. Whether we are to say that the passionate quest of the occult has been prolific in mental disturbances, or whether we had better say that persons with ill-balanced minds take especial delight in the search for the occult, the practical result is about the same. The impelling motive is not very different from that of the circle-squarers ; it is pleasing to one's self-love to feel that one discerns things to which all other people are blind. Hence the number of mare's-nests that have been complacently stared into by learned donkeys is legion. Mere erudition is no sure safeguard against the subtle forms which the temptation takes on, as we may see from the ingenuity that has been wasted on the Great Pyramid. In 1864, Piazzi Smyth, Astronomer Royal for Scotland, published his book entitled " Our Inheritance in the Great Pyramid," and afterward followed it with other similar books. Whatever may have been the original complexion of this gentleman's mind, it was not such as to prevent his attaining distinction and achieving usefulness as a practical astronomer. But the pyramids were too much for his mental equilibrium. As De Morgan kindly puts it, " his work on Egypt is paradox of a very high order, backed by a great quantity of useful labour, the results of which will be made available by those who do not receive the paradoxes."

The pyramidal tombs of Egyptian kings were an evolution in stone or brick from the tumulus of earth which in prehistoric ages was heaped over the body of the war chief. They are objects of rare dignity and interest, not only from their immense size, but from sundry peculiarities in their construction. In their orientation great care was taken, though usually with imperfect success. Their sides face the four cardinal points, and the descending entry-way forms a kind of telescope, from the bottom of which an observer, sixty centuries ago, could look out at what was then the polestar. These and other features of the pyramids are no doubt connected with Egyptian religion, and may very likely have subserved astrological purposes. But what say the pyramid cranks, or "pyramidalists," as they have been called?

According to them, the builders of the Great Pyramid were supernaturally instructed, probably by Melchizedek, King of Salem. Thus they were enabled to place it in latitude 30° N.; to make its four sides face the cardinal points; to adopt the sacred cubit, or one twenty millionth part of the earth's polar axis, as their unit of length; "and to make the side of the square base equal to just so many of these sacred cubits as there are days and parts of a day in a year. They were further by supernatural help enabled to square the circle, and symbolized their victory over this problem by making the pyramid's height bear to the perimeter of the base the ratio which the radius of a circle bears to the circumference." [1] In like manner, by immediate divine revelation, the builders of the pyramid were instructed as to the exact shape and density of the earth, the sun's distance, the precession of the equinoxes, etc., so that their figures on all these subjects were more accurate than any that modern science has obtained, and these figures they built into the pyramid. They also built into it the divinely revealed and everlasting standards of "length, area, capacity, weight, density, heat,

[1] Proctor, *The Great Pyramid*, p. 43.

time, and money," and finally they wrought into
its structure the precise date at which the millennium is to begin. All this valuable information, handed down directly from heaven, was thus
securely bottled up in the Great Pyramid for six
thousand years or so, awaiting the auspicious day
when Mr. Piazzi Smyth should come and draw the
cork. Why so much knowledge should have been
bestowed upon the architects of King Cheops, only
to be concealed from posterity, is a pertinent question; and one may also ask, why was it worth
while to bring a Piazzi Smyth into the world to
reveal it, since plodding human reason had after
all by slow degrees discovered every bit of it, except
the date of the millennium? Why, moreover, did
the revelation thus elaborately buried in or about
B. C. 4000 come just abreast of the scientific knowledge of A. D. 1864, and there stop short? Is it
credible that old Melchizedek knew nothing about
the telephone, or the Roentgen ray, or the cholera
bacillus? Our pyramidalists should be more enter-
prising, and elicit from their venerable fetish some
useful hints as to wireless telegraphy, or the ven-
tilation of Pullman cars, or the purification of
Pennsylvania politics. Perhaps the last-named
problem might vie in difficulty with squaring the
circle!

The lucubrations of Piazzi Smyth, like those of Miss Delia Bacon, called into existence a considerable quantity of eccentric literature. For example, there is Skinner's "Key to the Hebrew-Egyptian Mystery in the Source of Measures originating the British Inch and the Ancient Cubit," published in Cincinnati in 1875, a tall octavo of 324 pages, bristling with diagrams and decimals, Hebrew words and logarithms. The book begins by getting the circle neatly squared, and then goes on to aver that sundry crosses, including the Christian cross, are an emblematic display of the origin of measures. The "mound-builders" come in for a share of the author's attention; for the mounds are "alike Typhonic emblems with the pyramid of Egypt and with Hebrew symbols." A Typhonic emblem relates to Typhon, the "lord of sepulture," whose Egyptian representative was the crocodile, as his Hebrew representative was the hog; "exemplified in the Christian books by the devil leaving the man and passing into the herd of swine, which thereupon rushed into the sea, another emblem of Typhon." Yet another such emblem is a mound in Ohio which simulates the contour of an alligator. A certain Aztec pyramid, described by Humboldt, has 318 niches, apparently in allusion to the days of the old Mexican civil calendar. Mr. Skinner

sees in this numeral the value of **Pi**, and furthermore informs us that 318 is the Gnostic symbol for
Christ, as well as the number of Abraham's trained
servants. Frequent use of it is made in the Great
Pyramid; for example, multiplied by six it gives
the height of the king's chamber, and multiplied
by two it gives half the base side of that apart-
ment. Our author then puts the pyramid into a
sphere, and after this feat it is an easy transition
to Noah's flood, the zodiac, and modern ritualism.
Of similar purport, though more concise than this
octavo, is Dr. Watson Quinby's "Solomon's Seal,
a Key to the Pyramid," published at Wilmington,
Delaware, in 1880. From this little book we learn
that " in the early days of the world some one measured the earth, and found its diameter, in round
numbers, to be 41,569,000 feet, or 498,828,000
inches; " also that " Vishnu means Fish - Nuh,
Noah-the-Fish, in allusion to his sojourn in the
ark." Moreover, the Institutes of Manu were
written by Noah, since Maha-Nuh = Great-Noah!
With equal felicity, Rev. Edward Dingle (in his
" The Balance of Physics, the Square of the Circle, and the Earth's True Solar and Lunar Distances," London, 1885, pp. 246) declares that
" my success, let it be held what it may, was
secured by cleaving to the Mosaic initiation of the

Sabbatic number for my radius." At the end of his book Mr. Dingle exclaims : " To the Lord be all thanksgiving, who has kept my intellect and the directing of its thoughts sound, while seeking to deliver his word from the exulting shouts of his enemies and the seducers of mankind ! "

From these grotesque rigmaroles it is not a long step to the lucubrations of the writers in whose bonnets the bee of prophecy has buzzed until they have come to fancy themselves skilled interpreters. There is apt to be the same droll mixing of arithmetic with history that we find among the pyramid cranks, and to the performance of such antics the book of Daniel and the Apocalypse present irresistible temptations. In my library days, I never used to pick up a commentary on either of those books without looking for some of the stigmata or witch-marks of crankery. Many a feeble intellect has been toppled over by that shining image, with head of gold and feet of iron and clay, which Nebuchadnezzar beheld in a dream. For example, let us take a few sentences from " Emmanuel : An Original and Exhaustive Commentary on Creation and Providence Alike. By an Octogenarian Layman," London, 1883, pp. 420 : " Upwards of thirty years ago, a fancy for chronological research, fostered by boundless leisure and a competent facility in mental calculation, riveted my attention on the metallic image, in the vague hope of symmetrizing the four sections of the collective emblem with the successive dominations of the individual empires. Failing in so shadowy an aspiration, I seemed to be more than compensated by detecting an identity of duration, equally pregnant and positive, between the gold and the silver and the brass and the iron taken together on the one hand, and the mountain that was to crush them all to powder on the other, — the former aggregate being assumed to stretch from Nebuchadnezzar's succession in 606 B. C. to the dethronement of Augustulus in 476 A. D., and the latter again from the epoch just specified to Elizabeth's purgation of the Sanctuary in 1558." Having thus taken two equal periods of 1082 years, our Octogenarian proceeds to break them up (Heaven knows why!) each into four periods of 68, 204, 269, and 541 years. Then we are treated to the following equations : —

$$68 = 2 \times 34$$
$$204 = 6 \times 34$$
$$269 = 5 \times 34 + 3 \times 33$$
$$541 = 13 \times 34 + 3 \times 33$$

Hence, "with such a fulcrum as the Lamb slain before the foundation of the world, and such a lever as the span of the Victim's sublunary humiliation, was I too rash in aiming at a result infinitely grander than Archimedes's speculative displacement of the earth ? "

That eminent mathematician, Dr. Nathaniel Bowditch, used to say that sometimes, when Laplace passed from one equation to the next with an "evidently," he would find a week's study necessary to cross the abyss which the transcendent mind of the master traversed in a single leap. I fancy that more than a week would be needed to fathom the Octogenarian's "hence," and it would by no means be worth while to go through so much and get so little. After a few pages of the Octogenarian, we are prepared to hear that in 1750 one Henry Sullamar squared the circle by the number of the Beast with seven heads and ten horns; and that in 1753 a certain French officer, M. de Causans, " cut a circular piece of turf, squared it, and deduced original sin and the Trinity." [1]

The reader is doubtless by this time weary of so much tomfoolery ; but as it is needful, for the due comprehension of crankery and its crotchets, that he should by and by have still more of it, I will give him a moment's relief while I tell of a little game with which De Morgan and Whewell once

[1] De Morgan, p. 179.

amused themselves. The task was to make a sentence which should contain all the letters of the alphabet, and each only once. "No one," says De Morgan, "has done it with *v* and *j* treated as consonants; but *you* and *I* can do it" (*u* and *i :* oh, monstrous pun !). Dr. Whewell got only separate words, and failed to make a sentence : *phiz, styx, wrong, buck, flame, quid.* Very pretty, but De Morgan beat him out of sight with this weird sentiment ; *I, quartz pyx, who fling muck beds!* Well, what in the world can that mean ? "I long thought that no human being could say it under any circumstances. At last I happened to be reading a religious writer — as he thought himself — who threw aspersions on his opponents thick and threefold. Heyday! came into my head, this fellow flings muck beds : he must be a quartz pyx. And then I remembered that a pyx is a sacred vessel, and quartz is a hard stone, as hard as the heart of a religious foe-curser. So that the line is the motto of the ferocious sectarian, who turns his religious vessels into mud-holders for the benefit of those who will not see what he sees." [1]

I cite this drollery to show the world-wide difference between the playful nonsense of the wise man and the strenuous nonsense of the mono-

[1] De Morgan, p. 163.

maniac; in this little *cabbala alphabetica*, more-over, a great deal of the cabalistic lore which cumbers library shelves is neatly satirized.

As already observed, my rule was never to put into the class of eccentric literature any books save such as seemed to have emanated from diseased brains. To hold an absurd belief, to write in its defense, to shape one's career in accordance with it, is no proof of an unsound mind. Of the hundreds of enthusiasts who spent their lives in quest of the philosopher's stone, many were doubtless cranks; but many were able thinkers who made the best use they could of the scientific resources of their time. Wrong ways must often be tried before the right way can be found. Even the early circle-squarers cannot fairly be charged with crankery; they sinned against no light that was accessible to them. But anybody who to-day should advertise a recipe for turning base metals into gold would meet with a chill welcome from chemists. He would speedily be posted as a quack, though doubtless many weak heads would be turned by him. It is the perverse sinning against light that is one of the most abiding features of crankery, and from this point of view such a book as "Coin's Financial School" has many claims for admission to the limbo of eccentric literature.

About seventy years ago, one John Ranking published in London a volume entitled " Historical Researches on the Conquest of Peru, Mexico, Bogota, Natchez, and Talomeco,[1] in the Thirteenth Century, by the Mongols, accompanied with Elephants." It is well known that in 1281 the Mongols, after conquering pretty much everything from the Carpathian Mountains and the river Euphrates to the Yellow Sea, invaded Japan. A typhoon dispersed their fleet ; and their army of more than 100,000 men, cut off from its communications, was completely annihilated by the Japanese. But Mr. Ranking believed that this wholesale destruction was a fiction of the chroniclers. He maintained that most of the army escaped in a new fleet and crossed the Pacific Ocean, taking with them a host of elephants, with the aid of which they made extensive conquests in America and founded kingdoms in Mexico and Peru. The widespread fossil remains of the American mastodon he took to be the bones of these Mongolian elephants. Now, this is an extremely wild theory, unsound and untenable in every particular, but it does not bring Mr. Ranking's book within the class of eccentric literature. The author was deficient in scholarship and in critical judgment, but he was not daft.

[1] A site not far from that of Evansville, Indiana.

A very different verdict must be rendered in the
case of Mr. Edwin Johnson's book, called " The
Rise of Christendom," published in London in
1890, an octavo of 500 pages. According to Mr.
Johnson, the rise of Christendom began in the
twelfth century of our era, and it was preceded by
two centuries of Hebrew religion, which started in
Moslem Spain ! First came Islam, then Judaism,
then Christianity. The genesis of both the latter
was connected with that revolt against Islam which
we call the Crusades. What we suppose to be
the history of Israel, as well as that of the first
eleven Christian centuries, is a gigantic lie, con-
cocted in the thirteenth century by the monks of
St. Basil and St. Benedict. The Roman emperors
knew nothing of Christianity, and the multifarious
allusions to it in ancient writers are all explained
by Mr. Johnson as fraudulent interpolations. As
for the Greek and Latin fathers, they never existed. " The excellent stylist, who writes under
the name of Lactantius, not earlier than the fourteenth century ; " " the Augustinian of the fourteenth or fifteenth century, who writes the romantic Confessions," — such is the airy way in which
the matter is disposed of. As for the New Testament, " it is not yet clear whether the book was
first written in Latin or in Greek." This reminds

me of something once said by **Rev.** Robert Taylor,
a crazy clergyman who in 1827 suffered impris-
onment for blasphemy, and came to be known as
the Devil's Chaplain.   Taylor declared that for
the book of Revelation there was no Greek original at all, but Erasmus wrote it in Switzerland, in
the year 1516.   The audience, or part of it, probably took **Taylor's** word as sufficient; and in like
manner not a syllable of proof is alleged for Mr.
Johnson's prodigious assertions.   From cover to
cover, there is no trace of a consciousness that proof
is needed; it is simply, Thus saith **Edwin Johnson.**   The man who can write such a book is surely
incapable of making a valid will.

Another acute phase of insanity is exemplified
in Nason's " History of the Prehistoric Ages, written by the Ancient Historic Band of Spirits " (Chicago, 1880).   This is a mediumistic affair.   The
ancient band consists of four-and-twenty spirits,
the eldest of whom occupied a material body
46,000 years ago, and the youngest 3000 years
ago.   They dictated to Mr. Nason the narrative,
which begins with the origin of the solar system
and comes down to Romulus and Remus, betraying
on every page the preternatural dullness and ignorance so characteristic of all the spirits with whom
mediums have dealings.

Concerning the Bacon - Shakespeare lunacy a word must suffice. As I have shown in a previous essay, the doubt concerning the authorship of Shakespeare's plays was in part a reaction against the extravagances of doting commentators ; but in its original form it was simply an insane freak. The unfortunate lady who gave it currency belonged to a distinguished Connecticut family, and the story of her malady is a sad one. At the age of eight-and-forty she died in the asylum at Hartford, two years after the publication of her book, "The Philosophy of Shakespeare's Plays Unfolded." The suggestion of her illustrious namesake, and perhaps kinsman, as the author of Shakespeare's works, was a clear instance of the megalomania which is a well-known symptom of paranoia; and her book has all the hazy incoherence that is so quickly recognizable in the writings of the insane. A friend of mine once asked me if I did not find it hard to catch her meaning. "Meaning!" I exclaimed, "there's none to catch." Among the books of her followers are all degrees of eccentricity. That of Nathaniel Holmes stands upon the threshold of the limbo; while as for Ignatius Donnelly, all his works belong in its darkest recesses.

The considerations which would lead one to consign a book to that limbo are often complex.

There is Miss Marie Brown's book, " The Icelandic Discoverers of America; or, Honour to whom Honour is Due." In maintaining that Columbus knew all about the voyages of the Northmen to Vinland, and was helped thereby in finding his way to the Bahamas, there is nothing necessarily eccentric. Professor Rasmus Anderson has defended that thesis in a book which is able and scholarly, a book which every reader must treat with respect, even though he may not find its arguments convincing. But when Miss Brown declares that the papacy has been partner in a conspiracy for depriving the Scandinavians of the credit due them as discoverers of America, and assures us that this is a matter in which the interests of civil and religious liberty are at stake, one begins to taste the queer flavour ; and, taking this in connection with the atmosphere of rage which pervades the book, one feels inclined to place it in the limbo. For example : " What but Catholic genius, the genius for deceit, for trickery, for secrecy, for wicked and diabolical machinations, could have pursued such a system of fraud for centuries as the one now being exposed ! What but Catholic genius, a prolific genius for evil, would have attempted to rob the Norsemen of their fame, . . . and to foist a miserable Italian adventurer and

upstart upon Americans as the true candidate for
these posthumous honours, — the man or saint to
whom they are to do homage, and through this
homage allow the Church of Rome to slip the yoke
of spiritual subjection over their necks ! "

A shrill note of anger is sometimes the sure
ear-mark of a book from Queer Street. Anger is,
indeed, a kind of transient mania, and eccentric
literature is apt to be written in high dudgeon.
When you take up a pamphlet by " Vindex," and
read the title, " A Box on Both Ears to the Powers
that ought not to be at Washington," you may be
prepared to find incoherency. I once catalogued
an edition of Plutarch's little essay on Superstition,
and was about to let it go on its way, along with
ordinary Greek books, when my eye happened to
fall upon the last sentence of the editor's preface :
" I terminate this my Preface by consigning all
Greek Scholars to the special care of Beelzebub."
" Oho ! " I thought, " there 's a cloven foot here ;
perhaps, if we explore further, we may get a whiff
of brimstone." And it was so.

It thus appears that the topics treated in eccentric literature are numerous and manifold.
Not only, moreover, has this department its vigor-
ous prose-writers ; it has also its inspired poets.
Witness the following lines from the volume entitled " Eucleia " (Salem, 1861) : —

> " Hark, hear that distant **boo-oo-oo**,
>   **As,** walking by moonlight,
> He whistles, instructing Carlo
>   To be still, and not bite."

But even this lofty flight of inspiration is out-flown by Mr. John Landis, who was limner and draughtsman as well as poet. In his " Treatise on Magnifying God " (New York, 1843) he gives us an engraved portrait of himself surrounded by ministering angels, and accompanies it by an ode to himself, one verse of which will suffice : —

> " With Messrs. Milton, Watts, and Wesley,
>   Familiar thy Name will e'er be.
> Of America's Poets thou
>   Stand'st on the foremost list now ;
> On the pinions of fame does shine,
> *Landis !* brightened by ev'ry line,
> From thy poetic pen in rhyme,
> Thy name descends to the end of time."

Immortality of fame is something desired by many, but attained by few. Physical immortality is something which has hitherto been supposed to be inexorably denied to human beings. The phrase " All men are mortal " figures in text-books of logic as the truest of truisms. But we have lately been assured that this is a mistake. It is only an induction based upon simple enumeration, and the first man who escapes death will disprove it. So, at least, I was told by a very downright person who called on me some years ago with a huge parcel of manuscript, for which he wanted me to find him a publisher. He had been cruelly snubbed and ill-used, but truth would surely prevail over bigotry, as in Galileo's case. I took his address and let him leave his manuscript. Its recipe for physical immortality, diluted through 600 foolscap pages, was simply to learn how to go without food! Usually such a regimen will kill you by the fifth day, but if, at that critical moment, while at the point of death, you make one heroic effort and stay alive, why, then you will have overcome the King of Terrors once for all. I returned the gentleman's manuscript with a polite note, regretting that his line of research was so remote from those to which I was accustomed that I could not give him intelligent aid.

On one of the beautiful hills of Petersham, near the centre of Massachusetts, there dwelt a few years since a small religious community of persons who believed that they were destined to escape death. Not science, but faith, had won for them this boon. They believed that the third person of the Trinity was incarnated in their leader or high priest, Father Howland. This community, I believe, came from Rhode Island about forty years ago, and at the height of its prosperity may have

numbered twenty-five or thirty men and women.
Their establishment consisted of one large man-
sard-roofed house, with barns and sheds and a
good-sized farm. Their housekeeping was tidy;
and they put up apple-sauce. They maintained
that the eighteen and a half centuries of the so-
called Christian era have really been the dispensation of John the Baptist, and that the true
Christian era was ushered in by the Holy Ghost
in the person of Father Howland, through believing in whom Christians might attain to eternal life
on this planet. They had their Sabbath on Saturday, and worked in the fields on Sunday; and
they made sundry distinctions between clean and
unclean foods, based upon their slender understanding of the Old Testament.

For a few years these worthy people enjoyed
the simple rural life on their pleasant hillside without having their dream of immortality rudely
tested. When one member fell ill and died, and
was presently followed by another, it was easy to
dispose of such cases by asserting that the de-
ceased were not true believers; they were black
sheep, hypocrites, pretenders, whited sepulchres,
and their deaths had purified the flock. But the
next one to die was Father Howland himself. On
a warm summer day of 1874, as he was driving in

his buggy over a steep mountain road, the horse
shied so violently as to throw out the venerable
sage against a wood-pile, whereupon sundry loose
logs fell upon his head and shoulders, inflicting
fatal wounds. Then a note of consternation mingled with the genuine mourning of the little community. It was a perplexing providence. About
twelve months afterward I made my first visit
to these people, in company with my friend Dr.
William James and five carriage-loads of city folk
who were spending the summer at Petersham. It
was a Saturday morning, and all the worshippers
were in their best clothes. They received us with
a quiet but cordial welcome, and showed us into
a spacious parlour that was simply brilliant with
cheerfulness. Its west windows looked down upon
a vast and varied landscape, with rich pastures,
smiling cornfields, and long stretches of pine forest
covering range upon range of hills moulded in
forms of exquisite beauty. Beyond the foreground
of delicate yellow and soft green tints the eye rested
upon the sombre green of the woodland, and behind it all came the rich purple of the distant hills,
fitfully checkered with shadows from the golden
clouds. Here and there gleamed the white church
spires of some secluded hamlet, while on the horizon, seventy miles distant, arose the lofty peak of

old Greylock. Thence to Mount Grace, in one huge sweep, the entire breadth of Vermont was displayed, a wilderness of pale-blue summits blending with the sky; and over all, and part of it all, was the radiant glory of the September sunshine.

"Truly," said I to one of the brethren, a man of saintly face, "if you are expecting to dwell forever upon the earth, you could not have chosen a more inspiring and delightful spot." "Yes, indeed," he replied, "it seems too beautiful to leave." The topic which agitated the little community was thus brought up for discussion, and, except for a brief prayer, the ordinary Sabbath exercises were set aside for this purpose. All these people seemed polite and gentle in manner; their simple-mindedness was noticeable, and their ignorance was abysmal, though I believe they could all read the Bible and do a little writing and arithmetic. In the facial expression of every one I thought I could see something that betrayed more or less of a lapse from complete sanity. Only one of the whole number showed any sense of humour, a keen-eyed old woman, yclept Sister Caroline, who could argue neatly and make quaint retorts. She and the man of saintly face were the only interesting personalities; the rest were but soulless clods.

It soon appeared that the belief in terrestrial

immortality had not yet been seriously shaken by Father Howland's demise. There were some curious incipient symptoms of a resurrection myth. Their leader's death had been heralded by signs and portents. One aged brother, while taking his afternoon nap in a rocking-chair, fell forward upon the floor, bringing down the chair upon his back; and at that identical moment another brother rushed in from the garden, exclaiming, " I have seen with these eyes the glory of the Lord revealed!" Evidently, the fall of the rocking-chair prefigured the fall of the wood-pile, and the moment of Howland's fatal injury was the moment of his glorification. Then it was remembered by Sister Caroline and others that he had lately foretold his apparent death, and declared that it was to be only an appearance. " Though I shall seem to be dead, it will only be for a little while, and then I shall return to you."

The morning's conversation made it clear that these simple folk were unanimous in believing that the completion of Father Howland's work demanded his presence for a short time in the other world, and that he would within a few more weeks or months return to them. It seemed to Dr. James and myself that the conditions were favourable to the sudden growth of a belief in his resurrection,

and for some time after that visit we half expected
to hear that one or more of the household had seen
him. In this, however, we were disappointed. I
suspect that its mental soil may, after all, have
been too barren for such a growth.

Seven years elapsed before my second and last
visit to these worthy people. In the mean time a
large addition had been made to the principal
house, nearly doubling its capacity; and I was told
that the community had been legally incorporated
under the Hebrew title of Adoni-shomo, or " The
Lord is there." One would naturally infer that
the membership had increased, but the true explanation was very different. On a Saturday afternoon in the summer of 1882, in company with
fifteen friends, I visited the community. Our reception this time was something more than polite;
there was a noticeable warmth of welcome about
it. We were ushered into one of the newly built
rooms, — a long chapel, with seats on either side
and a reading-desk at one end. All the women,
both hosts and guests, took their seats on one side,
all the men on the other. A whisper from my
neighbour informed me that the community was
reduced to twelve persons : thus the guests out-
numbered the hosts. The high priest, Father Richards, a venerable man of ruddy hue, with enormous

beard as white as snow, stood by the reading-desk, and in broken tones gave thanks to God, while abundant tears coursed down his cheeks. Now, he said, at last the word of the Lord was fulfilled. Two or three years ago the word had come that they must build a chapel and add to their living-rooms, for they were about to receive a large accession of new converts. So — just think of it, gentle reader, in the last quarter of this skeptical century — there was faith enough on that rugged mountain-side to put three or four thousand dollars, earned with pork and apple-sauce, into solid masonry and timber-work! And now at last, said Father Richards, in the arrival of this goodly company the word of the Lord was fulfilled! It seemed cruel to disturb such jubilant assurance, but we soon found that we need not worry ourselves on that score. The old man's faith was a rock on which unwelcome facts were quickly wrecked. Though we took pains to make it clear that we had only come for a visit, it was equally clear to him that we were to be converted that very afternoon, and would soon come to abide with the Adoni-shomo.

Then Sister Caroline, stepping forward, made a long metaphysical harangue, at the close of which she walked up one side of the room and down the

other, taking each person by the hand and saying
to each a few words. When she came to me she
suddenly broke out with a stream of gibberish, and
went on for five mortal minutes, pouring it forth
as glibly as if it had been her mother tongue.
After the meeting had broken up, I was informed
that this " speaking with tongues " was not uncommon with the Adoni-shomo. A wicked wag in our
party then asked Sister Caroline if she knew what
language it was in which she had addressed me.
" No, sir," she replied, " nor do I know the meaning of what I said : I only uttered what the Lord
put into my mouth." " Well," said this graceless
scoffer, with face as sober as a deacon's, " I am
thoroughly familiar with Hebrew, and I recognized
at once the very dialect of Galilee as spoken when
our Saviour was on the earth ! " At this, I need
hardly add, Sister Caroline was highly pleased.

By this time there had been so many deaths that
induction by simple enumeration was getting to be
too much for the Adoni-shomo. They were beginning to realize the old Scotchman's conception of
the elect : " Eh, Jamie! hoo mony d' ye thank
there be of the elact noo alive on earth ? " " Eh!
mabbee a doozen." " Hoot, mon, nae sae mony as
thot ! " We found our worthy hosts less willing
than of old to discuss their doctrine of terrestrial

immortality, and there were symptoms of a tendency to give it a Pickwickian construction. Since
that day, their little community has vanished, and
its glorious landscape knows it no more.

It is a pity that before the end it should not
have had a visit from Mr. Hyland C. Kirk, whose
book on " The Possibility of Not Dying " was published in New York in 1883. In this book the
philosophic plausibleness of the opinion that a time
will come when we shall no longer need to shuffle
off this mortal coil is argued at some length, but
the question as to how this is to happen is ignored.
Mr. Isaac Jennings, in his " Tree of Life " (1867),
thinks it can be accomplished by total abstinence
from " alcohol, tobacco, coffee, tea, animal food,
spices, and caraway." This is sufficiently specific ;
but Mr. Kirk's treatment of the question is so hazy
as to suggest the suspicion that he has nothing to
offer us.

I once knew such a case of a delusion without
any theory, or, if you please, the grin without the
Cheshire cat. In the course of a lecturing journey, some thirty years ago, I was approached by a
refined and cultivated gentleman, who imparted to
me in strict confidence and with much modesty of
manner the fact that he had arrived at a complete
refutation of the undulatory theory of light! To

ask him for some statement of his own theory was but ordinary courtesy; but whenever we arrived at this point — which happened perhaps half a dozen times — he would put on a smile of mystery and decline to pursue the subject. I assured him that he need have no fear of my stealing his thunder, for I had not the requisite knowledge; but he grew more darkly mysterious than ever, and said that the time for him to speak had not yet come.

A few months later, this gentleman, whom I will designate as Mr. Flighty, appeared in Cambridge, and came to my desk in the college library. Distress was written in his face. He had called upon Professor Silliman and other professors in Eastern colleges, and had been shabbily treated. Nobody had shown him any politeness except Professor Youmans, in whom he believed he had found a convert. "Ah!" I exclaimed, "then you told him your theory; perhaps the time has come when you can tell it to me." But no; again came the subtle smile, and he began to descant upon the persecution of Galileo, a favourite topic with cranks of all sorts. He asked me for some of the best books on the undulatory theory, and I gave him Cauchy, whereat he stood aghast, and said the book was full of mathematics which he could not read; but he would like to see Newton's Opticks, for that book did not uphold the undulatory theory. "Oh!" said I, "then are you falling back on the corpuscular theory?" "No, indeed; mine is neither the one nor the other," and again came the Sibylline smile. As I went for the book, I found Professor Lovering in the alcove, halfway up a tall ladder. "Hallo!" said I *sotto voce.* "There is a man in here who has upset the undulatory theory of light; do you want to see him?" "Heavens, no! Can't you inveigle him into some dark corner while I run away?" "Don't worry," I replied, — "make yourself comfortable; I'll keep him from you." So I lured Mr. Flighty into a discourse on the bigotry of scientific folk, while Old Joe, whose fears were not so easily allayed, soon stealthily emerged from his alcove and hurried from the hall.

The next time that I happened to be in New York, chatting with Youmans at the Century Club, I alluded to Mr. Flighty, who believed he had made a convert of him.

"Ay, ay," rejoined Youmans, "and he said the same of you."

"Indeed! Well, I suspected as much. Unless you drive a crank from the room with cuffs and jeers, he is sure to think you agree with him. I do not yet know what Mr. Flighty's theory is."

" Nor I," said Youmans.

" Do you believe he has any theory at all ? "

" Not a bit of it. He is a madman, and his belief that he has a theory is simply the form which his delusion takes."

" Exactly so," I said ; and so it proved. Severe business troubles had wrecked Mr. Flighty's mind, and it was not long before we heard that he had killed himself in a fit of acute mania.

My story must not end with such a gruesome affair. Out of the many queer people I have known, let me mention one who is associated with pleasant memories of childhood and youth. This man was no charlatan, but a learned naturalist, of solid and genuine scientific attainments, who came to be a little daft in his old age. Dr. Joseph Barratt, whose life extended over three fourths of the present century, was born in England. He was at one time a pupil of Cuvier, and cherished his memory with the idolatrous affection which that wonderful man seems always to have inspired. Dr. Barratt, as a physician practising in Middletown, Connecticut, is one of the earliest figures in my memory, — a quaint and lovable figure. His attainments in botany and comparative anatomy were extensive ; he was more or less of a geologist, and well read withal in history and general litera-

ture, besides being a fair linguist. Though eminently susceptible of 'the tender passion, he never
married; he was neither a householder nor an
autocrat of the breakfast table, but dwelt hermit-
like in a queer snuggery over somebody's shop.
His working-room was a rare sight; so much confusion has not been seen since this fair world
weltered in its primeval chaos. With its cases
of mineral and botanical specimens, stuffed birds
and skeletons galore; with its beetles and spiders
mounted on pins, its brains of divers creatures in
jars of alcohol, its weird retorts and crucibles, its
microscopes and surgeon's tools, its shelves of mysterious liquids in vials, its slabs of Portland sand-
stone bearing footprints of Triassic dinosaurs, and
near the door a grim pterodactyl keeping guard
over all, it might have been the necromancing
den of a Sidrophel. Maps and crayon sketches,
mingled with femurs and vertebræ, sprawled over
tables and sofas and cumbered the chairs, till there
was scarcely a place to sit down, while everywhere in direst helter-skelter yawned and toppled
the books. And such books! There I first
browsed in Geoffroy St. Hilaire and Lamarck and
Blainville, and passed enchanted hours with the
" Règne Animal." The doctor was a courtly gentleman of the old stripe, and never did he clear a

chair for me without an apology, saying that he only awaited a leisure day to put all things in strictest order. Dear soul! that day never came.

Dr. Barratt was of course intensely interested in the Portland quarries, and they furnished the theme of the monomania which overtook him at about his sixtieth year. He accepted with enthusiasm the geological proofs of the antiquity of man in Europe, and presently undertook to reinforce them by proofs of his own gathering in the Connecticut Valley. An initial difficulty confronted him. The red freestone of that region belongs to the Triassic period, the oldest of the secondary series. It was an age of giant reptiles, contemporary with the earliest specimens of mammalian life, and not a likely place in which to look for relics of the highest of mammals. But Dr. Barratt insisted that this freestone is Eocene, thus bringing it into the tertiary series; and while geologists in general were unwilling to admit the existence of man before the Pleistocene period, he boldly carried it back to the Eocene. Thus, by adding a few million years to the antiquity of mankind and subtracting a few million from that of the rocks, he was enabled at once to maintain that he had discovered in the Portland freestone the indisputable remains of an ancient human being with only

three fingers, upon whom he bestowed the name of
Homo tridactylus. For companions he gave this
personage four species of kangaroo, and from that
time forth discoveries multiplied.

Such claims, when presented before learned
societies with the doctor's quaint enthusiasm, and
illustrated by his marvellous crayon sketches,
were greeted with shouts of laughter. Among the
geologists who chiefly provoked his wrath was the
celebrated student of fossil footprints, Dr. Edward
Hitchcock. " Why, sir," he would exclaim, " Dr.
Hitchcock is a perfect fool, sir ! I can teach ten
of him, sir ! " In spite of all scoffs and rebuffs,
the old gentleman moved on to the end serene in
his unshakable convictions. A courteous listener
was, of course, a rare boon to him ; and so, in that
little town, it became his habit to confide his new
discoveries to me. When I was out walking, if
chary of my half hours (as sometimes happened),
a long detour would be necessary, to avoid his
accustomed haunts ; and once, on my return from
a journey, I had hardly rung the doorbell when he
appeared on the veranda with an essay entitled
" An Eocene Picnic," which he hoped to publish
in " The Atlantic Monthly," and which he insisted
upon reading to me then and there. At one time
a very large bone was found in one of the quarries,

which was pronounced by Dr. Hitchcock to have belonged to an extinct batrachian; but Dr. Barratt saw in it the bone of a pachyderm. "Why, sir," said he, "it was their principal beast of burden, — as big as a rhinoceros and as gentle as a lamb. The children of Homo tridactylus used to play about his feet, sir, in perfect safety. I call him *Mega-ergaton docile*, 'the teachable great-worker.' Liddell and Scott give only the masculine, *ergates*, but for a beast of burden, sir, I prefer the neuter form. A gigantic pachyderm, sir; and Dr. Hitchcock, sir, perfect fool, sir, says it was a bullfrog!"

The mortal remains of this gentle palæontologist rest in the beautiful Indian Hill Cemetery at Middletown, and his gravestone, designed and placed there by my dear friend, the late Charles Browning, is appropriate and noble. For the doctor was after all a sterling man, whose unobtrusive merits were great, while his foibles were not important. The stone is a piece of fossil tree-trunk, brought over from Portland, imbedded in an amorphous block untouched by chisel, save where, on a bit of polished surface, one reads the name and dates, with the simple legend, "The Testimony of the Rocks."

November, 1898.

NOTE

AN ACCOUNT OF THE ADONI-SHOMO COMMUNITY

From the *Springfield Republican*. (1876.)

As queer a people as are often met, and apparently as
upright and religious, withal, are the Community situated on the stage-road between Athol and Petersham,
and commonly known thereabouts as "Howlandites"
or "Fullerites." According to their account, nearly
twenty-one years ago, two Worcester women, Mrs. Sa-
rah J. Hervey and her sister, Caroline E. Hawks, had
come to hope for a divine revelation to them, and in ex-
pectation of it had gone to a camp-meeting at Groton.
Entering the meeting they heard a stranger " talking
in tongues," who proved to be the man to meet their
wants, in the person of Frederick T. Howland, a Quaker, of good social standing, from New Bedford. That
day, September 15, 1855, was the origin "in the faith,"
though not in temporal association, of the Community,
these three being the "pioneers," as Sister Hervey takes
pride in calling herself and associates. Mrs. Hervey's
husband died a year or two later, though not in the
faith, " these things," as they say, " having been beyond
him." Soon after, the new belief received the addition of eight persons from Athol, among them Leonard
C. Fuller, the present Spiritual head of the Community,

and his wife. In May, 1861, having been "moved by the Spirit" to form an association for living together, they settled at Fuller's, at the south end of Pleasant Street in Athol. In August, 1864, they removed to their present farm in Petersham. Brother Howland held the position of head of the body till killed by a runaway horse, not quite two years ago. His people considered him a prophet, and say the Lord spoke by him, and that he led them as Moses led the people of Israel.

Their religious belief in many respects resembles that of the Adventists, but differs in the vital point, that the reign of Christ, under the expected new dispensation, is to be spiritual, and not personal, as the Adventists hold. They construe the saying of John the Revelator, "I was in the Spirit on the Lord's day," to refer to a period of time to begin with the 7000th year of the world, which is near at hand. The judgment day they believe has already begun, and in a short time, at the opening of the new dispensation, the holy dead are to be raised. When a man who has only received "common" salvation dies, he has no consciousness till the resurrection; but some, who are "specially" saved, will not die. Miracles will be performed commonly. When the new dispensation begins they are to be of the 144,000 spoken of by John, and are to judge the nations. They do not believe in a hypothetical heaven somewhere in space; the earth is not to be destroyed but changed; and finally the devil is to be bound for a thousand years. They entirely denounce Spiritualism, saying that it is from the Devil, or Antichrist. Brother Howland, they say, lay down to rise with the prophets, and they have written out what they claim to be prophecies made by him months or years before his death as to the manner in which it should occur, which, judged by the event, are certainly striking.

The Community live mostly upon farinaceous food; they drink principally water, sometimes herb tea. No flesh is eaten, because there is to be a restitution of the order of things that prevailed in the garden of Eden, and nothing that grows in the ground, because the ground is cursed. They live on the apostolic plan of having all property in common. If any among them wish to get married, they have to leave the Community. Morning and evening they "wait before the Lord," standing, repeat the Lord's prayer, and read and explain the Bible, "as the Spirit gives utterance." Although the district public school is only a stone's throw away, the half-dozen children of the Community, whom they have adopted, "as the Lord sent them," are taught at home by Sister Hervey. Sometimes, the neighbours' children come in, also, and they are said to do better there than at the public school. The school gives an occasional visit before the family, and a Christmas tree is provided. No jewelry is worn, and they dress very plainly; though the "world's people" claim that the Community wear as expensive "fixin's" and show as much pride as they do. The Community observe a seventh-day Sabbath, extending from 6 P. M., Friday, to the same hour, Saturday. The exercises begin at 10 o'clock, Saturday, and continue without intermission till 3. They are of the opinion that they need not go to a synagogue or "where the minister has to go 'round and wake the people up, as he did down to the Advent Church in Athol, last Sunday." The family seat themselves in the parlour on three sides of the room, with the occasional visitors on the fourth side ; and the exercises consist of exhortations by the various members, according as they are moved by the Spirit, with abundant "amens" from the rest. If no one feels called upon to speak, they study the Bible. Often they break out into singing. The house is free to visitors at all times. Last year from June to October, they had over two hundred visitors, among them nineteen, unexpectedly, one Sabbath.[1]

Their number, now about twenty, varies from time to time. They say they do not expect additions, though recently they have received two or three which they count of considerable importance. One of them is a woman, formerly a member of the Shaker Community at Dayton, O., where she was not satisfied, who walked all the way from Ohio to join them ; another is an ex-Baptist minister from Athol. They say they have suffered considerable persecution "for righteousness' sake." Mrs. Fuller thinks she was cheated out of property which her mother left her, and, because of the faith, two of their number, while sick, they say, were turned out of

[1] This was my first visit, with Dr. James and other friends, as above described.

a house on School Street in this city. They add, however, that those forward in opposing them have died sudden or violent deaths. On the other hand, they are prospering; they own a farm of two hundred and ten acres, and Brother Asa Richards, their Temporal head, raises stock, grain, fruits, etc., nearly sufficient to support them. Brother Fuller, though their Spiritual head, [1] does the marketing, principally in Athol. They have decided to enlarge the house and build a chapel in a short time, " if the Lord permits." Last winter, to protect their property, they went to the secretary of the Commonwealth and were organized under recent state laws as a corporation, with all the powers of a chartered body, under the name of " Adoni-shomo," Hebrew for " the Lord is there ; " that name being found in Ezekiel xlviii. 35. All their property will now remain in the Community while a single member of it is living.

It may be added that the views which outsiders hold of their Community do not always agree with their own. A " brother " named Mann died, last fall, and, by their own confession, they had some difficulty with his heirs, but finally settled for a nominal sum. At first they refused to pay over anything, but the heirs, four in number, threatening law, they finally concluded that the Lord willed them to give up $800. The common belief is that Mann was worth as many thousands; at any

---

[1] Brother Fuller resigned in 1877, and was succeeded by Brother Richards as Spiritual head, or high priest of the Adoni-shomo.

rate, the Petersham property was deeded to him in connection with Howland. Athol people scout the idea that Howland had prophetic powers, and think that the Community was simply the result of a shrewd plan of his to get a living without working for it.

INDEX

INDEX

ELECTROTYPED AND PRINTED
BY H. O. HOUGHTON AND CO.

The Riverside Press

CAMBRIDGE, MASS., U. S. A.

www.ingramcontent.com/pod-product-compliance
Lightning Source LLC
Chambersburg PA
CBHW032014110726
47901CB00004B/1084